THE ANTI-GRAVITY FILES

Edited by
David Hatcher Childress

Other Books in the *Lost Science* Series:

•The Anti-Gravity Handbook
•Anti-Gravity & the World Grid
•Anti-Gravity & the Unified Field
•The Free-Energy Device Handbook
•The Time Travel Handbook
•Tapping the Zero Point Energy
•Quest for Zero Point Energy
•The Energy Grid
•The Bridge to Infinity
•The Harmonic Conquest of Space
•Anti-Gravity Propulsion Dynamics
•The Fantastic Inventions of Nikola Tesla
•The Cosmic Matrix
•The AT Factor
•Perpetual Motion
•Secrets of the Unified Field
•The Tesla Files

See more astounding science books at:
www.adventuresunlimitedpress.com

THE ANTI-GRAVITY FILES

Edited by
David Hatcher Childress

The Anti-Gravity Files
Edited by David Hatcher Childress

Adventures Unlimited Press

ISBN: 978-1-939149-75-6

Published by:
Adventures Unlimited Press
One Adventure Place, Box 74
Kempton, Illinois 60946 USA
auphq@frontiernet.net

www.adventuresunlimitedpress.com

Thanks to Tom Valone, Joel Dickinson, Robert Cook and many others.

Anti-gravity airships hover around a Tesla Tower.

A sea rescue using anti-gravity backpacks.

A firefighting crew with an anti-gravity backpack and platform for hoses.

Everything you can imagine is real.
—Pablo Picasso

There is something within me that might be illusion as it is often the case with young delighted people, but if I would be fortunate to achieve some of my ideals, it would be on the behalf of the whole of humanity.
—Nikola Tesla, 1892

TABLE OF CONTENTS

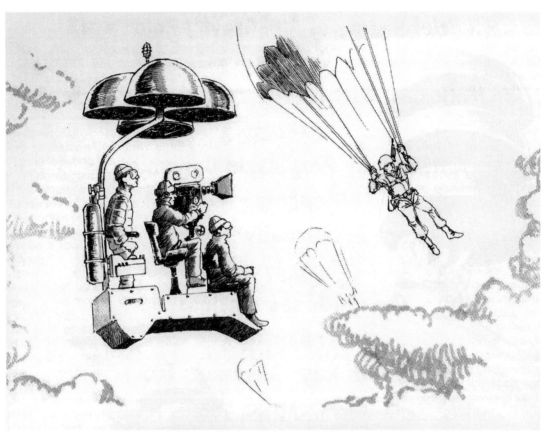

A crew with an anti-gravity platform filming a parachutist in action.

THE
ANTI-GRAVITY
FILES

Edited by
David Hatcher Childress

Chapter 1
A BRIEF HISTORY
OF ANTI-GRAVITY,
REACTIONLESS DRIVES &
FIELD PROPULSION

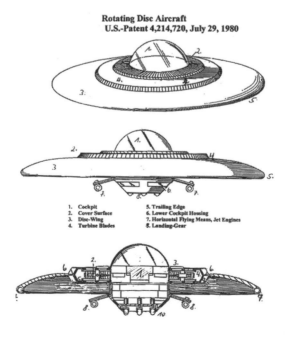

Rotating Disc Aircraft
U.S.-Patent 4,214,720, July 29, 1980

1. Cockpit
2. Cover Surface
3. Disc-Wing
4. Turbine Blades
5. Trailing Edge
6. Lower Cockpit Housing
7. Horizontal Flying Means, Jet Engines
8. Landing-Gear

Here we go with a brief history of anti-gravity as summed up by the modern day purveyor of all controversial things: Wikipedia. While hardly complete, it gives us the latest in mainstream thinking on such everyday things as field propulsion, gyros, reactionless drives and, yes, anti-gravity. While we are way over due for our anti-gravity belts, the state of anti-gravity research may be stronger than ever. Can anyone really define anti-gravity?
—David Hatcher Childress

Anti-Gravity (Wikipedia)

Anti-gravity also known as non gravitational field is an idea of creating a place or object that is free from the force of gravity. It does not refer to the lack of weight under gravity experienced in free fall or orbit, or to balancing the force of gravity with some other force, such as electromagnetism or aerodynamic lift. Anti-gravity is a recurring concept in science fiction, particularly in the context of spacecraft propulsion. Examples are the gravity blocking substance "Cavorite" in H. G. Wells' *The First Men in the Moon* and the Spindizzy machines in James Blish's *Cities in Flight*.

In Newton's law of universal gravitation, gravity was an external force transmitted by unknown means. In the 20th century, Newton's model was replaced by general relativity where gravity is not a force but the result of the geometry of spacetime.

Under general relativity, anti-gravity is impossible except under contrived circumstances. Quantum physicists have postulated the existence of gravitons, a set of massless elementary particles that transmit the force, and the possibility of creating or destroying these is unclear. "Anti-gravity" is often used colloquially to refer to devices that look as if they reverse gravity even though they operate through other means, such as lifters, which fly in the air by moving air with electromagnetic fields.

Hypothetical solutions

Gravity shields

In 1948 successful businessman Roger Babson (founder of Babson College) formed the Gravity Research Foundation to study ways to reduce the effects of gravity. Their efforts were initially somewhat "crankish," but they held occasional conferences that drew such people as Clarence Birdseye known for his frozen-food products and Igor Sikorsky, inventor of the helicopter. Over time the Foundation turned its attention away from trying to control gravity, to simply better understanding it. The Foundation nearly disappeared after Babson's death in 1967. However, it continues to run an essay award, offering prizes of up to $5,000. As of 2013, it is still administered out of Wellesley, Massachusetts, by George Rideout, Jr., son of the foundation's original director.

Winners include California astrophysicist George F. Smoot, who later won the 2006 Nobel Prize in physics. General relativity research in the 1950s

Main article: United States gravity control propulsion research

General relativity was introduced in the 1910s, but development of the theory was greatly slowed by a lack of suitable mathematical tools. Although it appeared that anti-gravity was outlawed under general relativity. It is claimed the US Air Force also ran a study effort throughout the 1950s and into the 1960s. Former Lieutenant Colonel Ansel Talbert wrote two series of newspaper articles claiming that most of the major aviation firms had started gravity control propulsion research in the 1950s. However, there is little outside confirmation of these stories, and since they take place in the midst of the policy by press release era, it is not clear how much weight these stories should be given. It is known that there were serious efforts underway at the Glenn L. Martin Company, who formed the Research Institute for Advance Study.

Major newspapers announced the contract that had been made between theoretical physicist Burkhard Heim and the Glenn L. Martin Company. Another effort in the private sector to master understanding of gravitation was the creation of the Institute for Field Physics, University of North Carolina at Chapel Hill in 1956, by Gravity Research Foundation trustee, Agnew H. Bahnson.

Military support for anti-gravity projects was terminated by the Mansfield Amendment of 1973, which restricted Department of Defense spending to only the areas of scientific research with explicit military applications. The Mansfield Amendment was passed specifically to end long-running projects that had little to show for their efforts. Under general relativity, gravity is the result of following spatial geometry (change in the normal shape of space) caused by local mass-energy. This theory holds that it is the altered shape of space, deformed by massive objects, that causes gravity, which is actually a property of deformed space rather than being a true force. Although the equations cannot normally produce a "negative geometry," it is possible to do so by using "negative mass." The same equations do not, of themselves, rule out the existence of negative mass. Both general relativity and Newtonian gravity appear to predict that negative mass would produce a repulsive gravitational field. In particular, Sir Hermann Bondi proposed in 1957 that negative gravitational mass, combined with negative inertial mass, would comply with the strong equivalence principle of general relativity theory and the Newtonian laws of conservation of linear momentum and energy.

Bondi's proof yielded singularity free solutions for the relativity equations. In July 1988, Robert L. Forward presented a paper at the AIAA/ASME/SAE/ASEE 24th Joint Propulsion Conference that proposed a Bondi negative gravitational mass propulsion system.

Bondi pointed out that a negative mass will fall toward (and not away from) "normal" matter, since although the gravitational force is repulsive, the negative mass (according to Newton's law, $F=ma$) responds by accelerating in the opposite of the direction of the force. Normal mass, on the other hand, will fall away from the negative matter. He noted that two identical masses, one positive and one negative, placed near each other will therefore self-accelerate in the direction of the line between them, with the negative mass chasing after the positive mass. Notice that because the negative mass acquires negative kinetic energy, the total energy of the accelerating masses remains at zero. Forward pointed out that the self-acceleration effect is due to the negative inertial mass, and could be seen induced without the gravitational forces between the particles.

The Standard Model of particle physics, which describes all presently known forms of matter, does not include negative mass. Although cosmological dark matter may consist of particles outside the Standard Model whose nature is unknown, their mass is ostensibly known—since they were postulated from their gravitational effects on surrounding objects, which implies their mass is positive. The proposed cosmological dark energy, on the other hand, is more complicated, since according to general relativity the effects of both its energy density and its negative pressure contribute to its gravitational effect.

Fifth force

Under general relativity any form of energy couples with spacetime to create the geometries that cause gravity. A longstanding question was whether or not these same equations applied to antimatter. The issue was considered solved in 1960 with the development of CPT symmetry, which demonstrated that antimatter follows the same laws of physics as "normal" matter, and therefore has positive energy content and also causes (and reacts to) gravity like normal matter (see gravitational interaction of antimatter). For much of the last quarter of the 20th century, the physics community was involved in attempts to produce a unified field theory, a single physical theory that explains the four fundamental forces: gravity, electromagnetism, and the strong and weak nuclear forces. Scientists have made progress in unifying the three quantum forces, but gravity has remained "the problem" in every attempt. This has not stopped any number of such attempts from being made, however.

Generally these attempts tried to "quantize gravity" by positing a particle, the graviton, that carried gravity in the same way that photons (light) carry electromagnetism. Simple attempts along this direction all failed, however, leading to more complex examples that attempted to account for these problems. Two of these, supersymmetry and the relativity related supergravity, both required the existence of an extremely weak "fifth force" carried by a graviphoton, which coupled together several "loose ends" in quantum field theory, in an organized manner. As a side effect, both theories also all but required that antimatter be affected by this fifth force in a way similar to anti-gravity, dictating repulsion away from mass. Several experiments were carried out in the 1990s to measure this effect, but none yielded positive results.

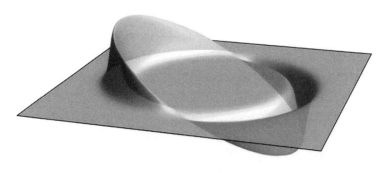

Two dimensional spacetime distortion induced by the Alcubierre metric.

In 2013 CERN looked for an antigravity effect in an experiment designed to study the energy levels within antihydrogen. The antigravity measurement was just an "interesting sideshow" and was inconclusive.

General-relativistic "warp drives"

There are solutions of the field equations of general relativity which describe "warp drives" (such as the Alcubierre metric) and stable, traversable wormholes. This by itself is not significant, since any spacetime geometry is a solution of the field equations for some configuration of the stress–energy tensor field (see exact solutions in general relativity). General relativity does not constrain the geometry of spacetime unless outside constraints are placed on the stress–energy tensor. Warp-drive and traversable-wormhole geometries are well-behaved in most areas, but require regions of exotic matter; thus they are excluded as solutions if the stress–energy tensor is limited to known forms of matter. Dark matter and dark energy are not understood enough at this present time to make general statements regarding their applicability to a warp-drive.

Breakthrough Propulsion Physics Program

During the close of the twentieth century NASA provided funding for the Breakthrough Propulsion Physics Program (BPP) from 1996 through 2002. This program studied a number of "far out" designs for space propulsion that were not receiving funding through normal university or commercial channels. Anti-gravity-like concepts were investigated under the name "diametric drive." The work of the BPP program continues in the independent, non-NASA affiliated Tau Zero Foundation.

Empirical claims and commercial efforts

There have been a number of attempts to build anti-gravity devices, and a small number of reports of anti-gravity-like effects in the scientific literature. None of the examples that follow are accepted as reproducible examples of anti-gravity.

Gyroscopic devices

Gyroscopes produce a force when twisted that operates "out of plane" and can appear to lift themselves against gravity. Although this force is well understood to be illusory, even under Newtonian models, it has nevertheless generated numerous claims of anti-gravity devices and any number of patented devices. None of these devices have ever been demonstrated to work under controlled conditions, and have often become the subject of conspiracy theories as a result. A famous example is that of Professor Eric Laithwaite of Imperial College, London, in the 1974 address to the Royal Institution.

The G-Engines Are Coming!

By far the most potent source of energy is gravity. Using it as power future aircraft will attain the speed of light.

A 60s newspaper cartoon about T. Townsend Brown.

Another "rotating device" example is shown in a series of patents granted to Henry Wallace between 1968 and 1974. His devices consist of rapidly spinning disks of brass, a material made up largely of elements with a total half-integer nuclear spin. He claimed that by rapidly rotating a disk of such material, the nuclear spin became aligned, and as a result created a "gravitomagnetic" field in a fashion similar to the magnetic field created by the Barnett effect. No independent testing or public demonstration of these devices is known.

In 1989, it was reported that a weight decreases along the axis of a right spinning gyroscope. A test of this claim a year later yielded null results. A recommendation was made to conduct further tests at a 1999 AIP conference.

Thomas Townsend Brown's gravitator

Further information: Biefeld-Brown effect, Electrogravitics, and United States gravity control propulsion research.

Brown's gravitator

In 1921, while still in high school, Thomas Townsend Brown found that a high-voltage Coolidge tube seemed to change mass depending on its orientation on a balance scale. Through the 1920s Brown developed this into devices that combined high voltages with materials with high dielectric constants (essentially large capacitors); he called such a device a "gravitator." Brown made the claim to observers and in the media that his experiments were showing anti-gravity effects. Brown would continue his work and produced a series of high-voltage devices in the following years in attempts to sell his ideas to aircraft companies and the military. He coined the names Biefeld–Brown effect and electrogravitics in conjunction with his devices. Brown tested his asymmetrical capacitor devices in a vacuum, supposedly showing it was not a more down to earth electrohydrodynamic effect generated by high voltage ion flow in air. Electrogravitics is a popular topic in ufology, anti-gravity, free energy, with government conspiracy theorists and related websites, in books and publications with claims that the technology became highly classified in the early 1960s and that it is used to power UFOs and the B-2 bomber. There is also research and videos on the internet purported to show lifter-style capacitor devices working in a vacuum, therefore not receiving propulsion from ion drift or ion wind being generated in air.

Follow-up studies on Brown's work and other claims have been conducted by R. L. Talley in a 1990 US Air Force study, NASA scientist Jonathan Campbell in a 2003 experiment, and Martin Tajmar in a 2004 paper. They have found that no thrust could be observed in a vacuum and that Brown's and other ion lifter devices produce thrust along their axis regardless of the direction of gravity consistent with electrohydrodynamic effects.

Gravitoelectric coupling

In 1992, the Russian researcher Eugene Podkletnov claimed to have discovered, whilst experimenting with superconductors, that a fast rotating superconductor reduces the gravitational effect. Many studies have attempted to reproduce Podkletnov's experiment, always to negative results.

Ning Li and Douglas Torr, of the University of Alabama in Huntsville proposed how a time dependent magnetic field could cause the spins of the lattice ions in a superconductor to generate detectable gravitomagnetic and gravitoelectric fields in a series of papers published between 1991 and 1993. In 1999, Li and her team appeared in Popular Mechanics, claiming to have constructed a working prototype to generate what she described as "AC Gravity." No further evidence of this prototype has been offered.

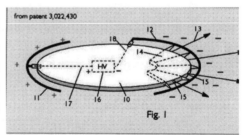

Portion of the 1960 patent by Brown.

Douglas Torr and Timir Datta were involved in the development of a "gravity generator" at the University of South Carolina. According to a leaked document from the Office of Technology Transfer at the University of South Carolina and confirmed to Wired reporter Charles Platt in 1998, the device would create a "force beam" in any desired direction and that the university planned to patent and license this device. No further information about this university research project or the "Gravity Generator" device was ever made public.

Ongoing research[citation needed] suggests that if antimatter indeed generates a -2.993% (consistent with astronomical observations negative gravitational effect then a device similar to Podkletnov's original experiment modified to direct positrons at the spinning disk could exert a weak effect on nearby objects in range. The effect would be centred on the spin axis and be intermittent due to positrons and electrons annihilating and heating up the lattice causing quenching. In order to see a significant effect the disk would need to be spun in a vacuum and held at its Tc due to the effect mentioned.

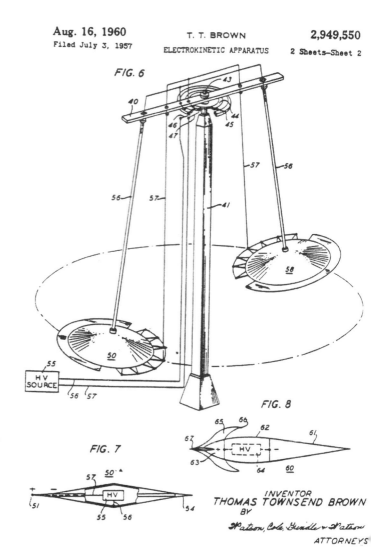

Aug. 16, 1960
Filed July 3, 1957

T. T. BROWN

ELECTROKINETIC APPARATUS

2,949,550

2 Sheets—Sheet 2

FIG. 6

FIG. 8

FIG. 7

HV SOURCE

INVENTOR
THOMAS TOWNSEND BROWN
BY
Watson, Cole, Grindle & Watson
ATTORNEYS

The 1960 patent for T. Townsend Brown.

Göde Award

The Institute for Gravity Research of the Göde Scientific Foundation has tried to reproduce many of the different experiments which claim any "anti-gravity" effects. All attempts by this group to observe an anti-gravity effect by reproducing past experiments have been unsuccessful thus far. The foundation has offered a reward of one million euros for a reproducible anti-gravity experiment.

Illustration: A "kinemassic field" generator from U.S. Patent 3,626,605: Method and apparatus for generating a secondary gravitational force field.

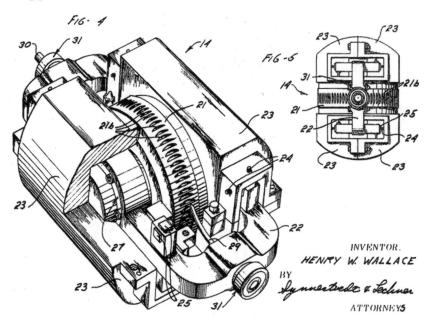

References

Peskin, M and Schroeder, D.; An Introduction to Quantum Field Theory (Westview Press, 1995) [ISBN 0-201-50397-2] Wald, Robert M. (1984). General Relativity. Chicago: University of Chicago Press. ISBN 0-226-87033-2.

Polchinski, Joseph (1998). String Theory, Cambridge University Press. A modern textbook Thompson, Clive (August 2003). "The Antigravity Underground." Wired. Archived from the original on 18 August 2010. Retrieved 23 July 2010. "On the Verge of Antigravity." About.com. Retrieved 23 July 2010.

Mooallem, J. (October 2007). "A curious attraction." Harper's Magazine. 315 (1889): 84–91. List of winners Goldberg, J. M. (1992). US air force support of general relativity: 1956–1972. In, J. Eisenstaedt & A. J. Kox (Ed.), Studies in the History of General Relativity, Volume 3 Boston, Massachusetts: Center for Einstein Studies. ISBN 0-8176-3479-7 Mallan, L. (1958). Space satellites (How to book 364). Greenwich, CT: Fawcett Publications, pp. 9–10, 137, 139. LCCN 58-001060 Clarke, A. C. (1957). "The conquest of gravity." Holiday. 22 (6): 62.

Bondi, H. (1957). "Negative mass in general relativity." Reviews of Modern Physics. 29 (3): 423–428. Bibcode:1957RvMP...29..423B. doi:10.1103/revmodphys.29.423. Forward, R. L. (1990, Jan.-Feb.), "Negative matter propulsion," Journal of Propulsion and Power, Vol. 6 (1), pp. 28–37; see also commentary Landis, G.A. (1991) "Comments on Negative Mass Propulsion," Journal of Propulsion and Power, Vol. 7, No. 2, p. 304. Supergravity and the Unification of the Laws of Physics, by Daniel Z. Freedman and Peter van Nieuwenhuizen, Scientific American, February 1978

Jason Palmer, Antigravity gets first test at Cern's Alpha experiment, bbc.co.uk, 30 April 2013 Tau Zero Foundation

"Eric LAITHWAITE Gyroscope Levitation." Rex research. rexresearch.com. Retrieved 23 October 2010. U.S. Patent 3,626,606 U.S. Patent 3,626,605 U.S. Patent 3,823,570

Hayasaka, H. & Takeuchi, S. (1989). "Anomalous weight reduction on a gyroscope's right rotations around the vertical axis on the Earth." Physics Review Letters. 63 (25): 2701–2704. Bibcode:1989PhRvL..63.2701H. doi:10.1103/PhysRevLett.63.2701.

Nitschke, J. M. & Wilmath, P. A. (1990). "Null result for the weight change of a spinning gyroscope." Physics Review Letters. 64 (18): 2115–2116. Bibcode:1989PhRvL..63.2701H. doi:10.1103/PhysRevLett.64.2115. Retrieved 5 January 2014.

Iwanaga, N. (1999). "Reviews of some field propulsion methods from the general relativistic standpoint." AIP Conference Proceedings. 458: 1015–1059.

Thompson, Clive (August 2003). "The Antigravity Underground." Wired Magazine. Thomas Valone, Electrogravitics II: Validating Reports on a New Propulsion Methodology, Integrity Research Institute, page 52-58

Thompson, Clive (August 2003). "The Antigravity Underground." Wired Magazine.

Tajmar, M. (2004). "Biefeld-Brown Effect: Misinterpretation of Corona Wind Phenomena." AIAA Journal. 42 (2): 315–318. Bibcode:2004AIAAJ..42..315T. doi:10.2514/1.9095.

Podkletnov, E; Nieminen, R (10 December 1992). "A possibility of gravitational force shielding by bulk YBa2Cu3O7−x superconductor." Physica C. 203 (3–4): 441–444. Bibcode:1992PhyC..203..441P. doi:10.1016/0921-4534(92)90055-H. Retrieved 29 April 2014.

N. Li; D. Noever; T. Robertson; R. Koczor; et al. (August 1997). "Static Test for a Gravitational Force Coupled to Type II YBCO Superconductors." Physica C. 281 (2-3): 260–267. Bibcode:1997PhyC..281..260L. doi:10.1016/S0921-4534(97)01462-7.

Woods, C., Cooke, S., Helme, J., and Caldwell, C., "Gravity Modification by High Temperature Superconductors," Joint Propulsion Conference, AIAA 2001–3363, (2001).

Hathaway, G., Cleveland, B., and Bao, Y., "Gravity Modification Experiment using a Rotating Superconducting Disc and Radio Frequency Fields," Physica C, 385, 488–500, (2003).

Tajmar, M., and de Matos, C.J., "Gravitomagnetic Field of a Rotating Superconductor and of a Rotating Superfluid," Physica C, 385(4), 551–554, (2003).

Li, Ning; Torr, DG (1 September 1992). "Gravitational effects on the magnetic attenuation of superconductors." Physical Review. B46: 5489–5495. Bibcode:1992PhRvB..46.5489L. doi:10.1103/PhysRevB.46.5489. Retrieved 6 March 2014.

Li, Ning; Torr, DG (15 January 1991). "Effects of a gravitomagnetic field on pure superconductors." Physical Review. D43: 457–459. Bibcode:1991PhRvD..43..457L. doi:10.1103/PhysRevD.43.457. Retrieved 6 March 2014.

Li, Ning; Torr, DG (August 1993). "Gravitoelectric-electric coupling via superconductivity." Foundations of Physics Letters. 6 (4): 371–383. Bibcode:1993FoPhL...6..371T. doi:10.1007/BF00665654. Retrieved 6 March 2014.

Wilson, Jim (1 October 2000). "Taming Gravity." Popular Mechanics. HighBeam Reseatch. Retrieved 5 January 2014.

Cain, Jeanette. "Gravity Conquered?." light-science.com. Retrieved 5 January 2014. "Patent and Copyright Committee List of Disclosures Reviewed Between July 1996 and June 1997 - USC ID." Retrieved 30 April 2014.

Platt, Charles (3 June 1998). "Breaking the Law of Gravity." Wired. Retrieved 1 May 2014. "The Göde award - One Million Euro to overcome gravity." Institute of Gravity Research. Retrieved 2 January 2014.

Reactionless Drive
From Wikipedia, the free encyclopedia
(Not to be confused with Field propulsion).

A reactionless drive is a device producing motion without the exhaust of a propellant. A propellantless drive is not necessarily reactionless when it constitutes an open system interacting with external fields; but a reactionless drive is a particular case of a propellantless drive as it is a closed system presumably in contradiction with the law of conservation of momentum and often considered similar to a perpetual motion machine. The name comes from Newton's third law, which is usually expressed as, "for every action, there is an equal and opposite reaction." A large number of infeasible devices, such as the Dean drive, are a staple of science fiction particularly for space propulsion. To date, no reactionless device has ever been validated under properly controlled conditions.

Closed systems

Through the years there have been numerous claims for functional reactionless drive designs using ordinary mechanics (i.e. devices not said to be based on quantum mechanics, relativity or atomic forces or effects). Two of these represent their general classes:

The "Dean drive" is perhaps the best known example of a "linear oscillating mechanism" reactionless drive; The "GIT" is perhaps the best known example of a "rotating mechanism" reactionless drive. These two also stand out as they both received much publicity from their promoters and the popular press in their day and both were eventually rejected when proven to not produce any reactionless drive forces. The rise and fall of these devices now serves as a cautionary tale for those making and reviewing similar claims.

Dean drive

The Dean drive was a mechanical device concept promoted by inventor Norman L. Dean. Dean claimed that his device was a "reactionless thruster" and that his working models could demonstrate this effect. He held several private demonstrations but never revealed the exact design of the models nor allowed independent analysis of them. Dean's claims of reactionless thrust generation were subsequently shown to be in error and the "thrust" producing the directional motion was likely to be caused by friction between the device and the surface on which the device was resting and would not work in free space.

Gyroscopic Inertial Thruster (GIT)

The Gyroscopic Inertial Thruster is a proposed reactionless drive based on the mechanical principles of a rotating mechanism. The concept involves various methods of leverage applied against the supports of a large gyroscope. The supposed operating principle of a GIT is a mass traveling around a circular trajectory at a variable speed. The high-speed part of the trajectory

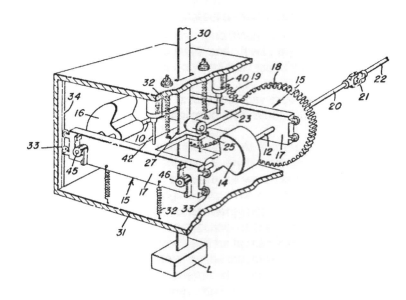

The Dean Drive.

allegedly generates greater centrifugal force than the low, so that there is a greater thrust in one direction than the other. Scottish inventor Sandy Kidd, a former RAF radar technician, investigated the possibility (without success) in the 1980s. He posited that a gyroscope set at various angles could provide a lifting force, defying gravity. In the 1990s, several people sent suggestions to the Space Exploration Outreach Program (SEOP) at NASA recommending that NASA study a gyroscopic inertial drive, especially the developments attributed to the American inventor Robert Cook and the Canadian inventor Roy Thornson. In the 1990s and 2000s, enthusiasts attempted the building and testing of GIT machines. Eric Laithwaite, the "Father of Maglev," received a US patent for his "Propulsion System," which was claimed to create a linear thrust through gyroscopic and inertial forces.

After years of theoretical analysis and laboratory testing of actual devices, no rotating (or any other) mechanical device has ever been found to produce unidirectional reactionless thrust in free space.

Open systems
Main article: Field propulsion Movement with thrust

Several kinds of thrust generating methods are in use or have been proposed that are propellantless, as they do not work like rockets and reaction mass is not carried nor expelled from the device. However they are not reactionless, as they constitute open systems interacting with electromagnetic waves or various kinds of fields. Most famous propellantless methods are the gravity assist maneuver or gravitational slingshot of a spacecraft accelerating at the expense of the momentum of the planet it orbits, through the gravitational field, or beam-powered propulsion using the radiation pressure of electromagnetic waves from a distant source like a laser. More speculative methods have also been proposed, like the Woodward effect, the quantum vacuum plasma thruster or various hypotheses trying to explain the thrust apparently produced by the EmDrive.

Movement without thrust

Because there is no well-defined "center of mass" in curved spacetime, general relativity allows a stationary object to, in a sense, "change its position" in a counter-intuitive manner, without violating conservation of momentum.

The Alcubierre drive is a hypothetical method of apparent faster-than-light propulsion for interstellar travel postulated from the theory of general relativity. Although this concept may be allowed by the currently accepted laws of physics, it remains unproven; implementation would require a negative energy density, and possibly a better understanding of quantum gravity. It is not clear how (or if) this effect could provide a useful means of accelerating an actual space vehicle and no practical designs have been proposed, but experiments are underway at NASA's Eagleworks Laboratories to attempt the first detection of an induced spacetime curvature, which could be the first step toward proving the validity of the concept.

A hypothetical "impulse engine" or "distortion engine" creating a gravitational potential in spacetime, with no negative energy source contrary to the Alcubierre drive, would also produce a reactionless motion, being a low velocity (non relativistic) warp drive.

Some bimetric theories of gravity with variable speed of light like the Janus cosmological model hypothesize apparent faster-than-light interstellar travel with no acceleration nor deceleration, using the energy difference of the two conjugated metrics to reach relativistic speeds after a mass inversion process.

"Swimming in spacetime" is a geometrical motive principle that exploits the curved spacetime metric of the gravitational field to permit an extended body undergoing specific deformations in shape, to change position. In weak gravitational fields, like that of Earth, the change in position per deformation cycle would be far too small to detect, but the concept remains of interest as the only unambiguous example of reactionless motion in mainstream physics.

Field Propulsion

Field propulsion is the concept of spacecraft propulsion where no propellant is necessary but instead momentum of the spacecraft is changed by an interaction of the spacecraft with external force fields, such as gravitational and magnetic fields from stars and planets. It is purely speculative and has not yet been demonstrated to be of practical use or theoretically valid.

Although not presently in wide use for space, there exist proven terrestrial examples of "Field Propulsion," in which electromagnetic fields act upon a conducting medium such as seawater or plasma for propulsion, is known as magnetohydrodynamics or MHD. MHD is similar in operation to electric motors, however rather than using moving parts or metal conductors, fluid or plasma conductors are employed. The EMS-1 and more recently the Yamato 1 are examples of such electromagnetic Field propulsion systems, first proposed in patent US 5333444. There is definitely potential to apply MHD to the space environment and experiments such as the NASA's electrodynamic tether, Lorentz Actuated Orbits, the wingless electromagnetic air vehicle, and magnetoplasmadynamic thruster (which does use propellant) lay a solid foundation for using "fields" to propel spacecraft without propellant and standard concepts of chemical thrust. Since electrodynamics is well proven science, electromagnetic fields themselves carry momentum (see the Nichols radiometer), and electromagnetic field propulsion is not limited to the ejection velocity of particle propellants these new concepts offer tremendous potential as a future space propulsion system. They represent a radical departure from current ideas of aeronautics and rocket propulsion, and as such are controversial, but field propulsion may offer the radical breakthroughs in performance capabilities required for deep space exploration. The main limiting factors appear to the generation of the significant amounts of electrical power required and a method of strongly coupling the fields to large volumes.

Electrohydrodynamics is another method whereby electrically charged fluids are used for propulsion and boundary layer control such as ion propulsion.

Other practical methods which could be loosely considered as field propulsion include: The gravity assist trajectory, which uses planetary gravity fields and orbital momentum; Solar sails and magnetic sails use respectively the radiation pressure and solar wind for spacecraft thrust; Aerobraking uses the atmosphere of a planet to change relative velocity of a spacecraft. The last two actually involve the exchange of momentum with physical particles and are not usually expressed as an interaction with fields, but they are sometimes included as examples of field propulsion since no spacecraft propellant is required.

Speculative methods

Other concepts that have been proposed are speculative, using "frontier physics" and concepts from modern physics. So far none of these methods have been unambiguously demonstrated, much less proven practical.

The Woodward effect is based on a controversial concept of inertia and certain solutions to the equations for General Relativity. Experiments attempting to conclusively demonstrate this effect have been conducted since the 1990s.

Although speculative, ideas such as coupling to the momentum flux of the zero-point electromagnetic wave field hypothesized in stochastic electrodynamics have a plausible basis for further investigation within the existing theoretical physics paradigm. In contrast, examples of proposals for field propulsion that rely on physics outside the present paradigms are various schemes for faster-than-light, warp drive and antigravity, and often amount to little more than catchy descriptive phrases, with no known physical basis. Until it is shown that the conservation of energy and momentum break down under certain conditions (or scales), any such schemes worthy of discussion must rely on energy and momentum transfer to the spacecraft from some external source such as a local force field, which in turn must obtain it from still other momentum and/or energy sources in the cosmos (in order to satisfy conservation of both energy and momentum).

Field propulsion based on physical structure of space

This concept is based on the general relativity theory and the quantum field theory from which the idea that space has a physical structure can be proposed. The macroscopic structure is described by the general relativity theory and the microscopic structure by the quantum field theory. The idea is to deform space around the space craft. By deforming the space it would be possible to create a region with higher pressure behind the space craft than before it. Due to the pressure gradient a force would be exerted on the space craft which in turn creates thrust for propulsion.

Due to the purely theoretical nature of this propulsion concept it is hard to determine the amount of thrust and the maximum velocity that could be achieved. Currently there are two different concepts for such a field propulsion system one that is purely based on the general relativity theory and one based on the quantum field theory.

In the general relativistic field propulsion system space is considered to be an elastic field similar to rubber which means that space itself can be treated as an infinite elastic body. If the space-time curves, a normal inwards surface stress is generated which serves as a pressure field. By creating a great number of those curve surfaces behind the space craft it is possible to achieve a unidirectional surface force which can be use for the acceleration of the space craft.

For the quantum field theoretical propulsion system it is assumed, as stated by the quantum field theory and quantum Electrodynamics, that the quantum vacuum consists out of a zero-radiating electromagnetic field in a non-radiating mode and at a zero-point energy state, the lowest possible energy state. It is also theorized that matter is composed out of elementary primary charged entities, partons, which are bound together as elementary oscillators. By applying an electromagnetic zero point field a Lorentz force is applied on the partons. Using this on a dielectric material could effect the inertia of the mass and that way create an acceleration of the material without creating stress or strain inside the material.

Conservation Laws

Conservation of momentum is a fundamental requirement of propulsion systems because in experiments momentum is always conserved, and is implicit in published work of Newton and Galileo. In each of the propulsion technologies, some form of energy exchange is required with momentum directed backward at light speed c or some lesser velocity v to balance the forward change of momentum. In absence of interaction with an external field, the power P that is required to create a thrust force F is given. $F = P/v$ when mass is ejected or $F = P/c$ if mass free energy is ejected. For a photon rocket the efficiency is too small to be competitive. Other technologies may have better efficiency if the ejection velocity is less than light speed, or a local field can interact with another large scale field of the same type residing in space, which is the intent of field effect propulsion.

Advantages

The main advantage of a field propulsion systems is that no propellant is needed, only an energy source. This means that no propellant has to be stored and transported with the space craft which makes it attractive for long term interplanetary or even interstellar manned missions. With current technology a large amount of fuel meant for the way back has to be brought to the destination which increases the payload of the overall space craft significantly. The increased payload of fuel, thus requires more force to accelerate it, requiring even more fuel which is the primary drawback of current rocket technology. Approximately 83% of a Hydrogen-Oxygen powered rocket, which can achieve orbit, is fuel.

See also
Abraham–Minkowski controversy
Beam-powered propulsion
Bernard Haisch
Field propulsion
Harold E. Puthoff
Inertialess drive
Perpetual motion
Spacecraft propulsion
Stochastic electrodynamics
RF resonant cavity thruster

References

Winchell D. Chung Jr. "Reactionless drives." Mills, Marc G.; Thomas, Nicholas E. (July 2006). Responding to Mechanical Antigravity (PDF). 42nd Joint Propulsion Conference and Exhibit. NASA. Archived from the original (PDF) on 2011-10-30.

"Engine With Built-in Wings." *Popular Mechanics*. Sep 1961.

"Detesters, Phasers and Dean Drives." *Analog*. June 1976.

Goswami, Amit (2000). The Physicists' View of Nature. Springer. p. 60. ISBN 0-306-46450-0.

LaViolette, Paul A. (2008). *Secrets of Antigravity Propulsion: Tesla, UFOs, and Classified Aerospace Technology*. Inner Traditions / Bear & Co. p. 384. ISBN 1-59143-078-X. New Scientist. 148: 96. 1995. Childress, David Hatcher (1990). *Anti-Gravity & the Unified Field*. Adventures Unlimited Press. p. 178. ISBN 0-932813-10-0. "The Adventures of the Gyroscopic Inertial Flight Team." 1998-08-13. U.S. Patent 5,860,317

Kakaes, Konstantin. "Warp Factor: A NASA scientist claims to be on the verge of faster-than-light travel: is he for real?, Popular Science, April 2013." PopSci.com. Retrieved 2014-11-22. http://ntrs.nasa.gov/archive/nasa/casi.ntrs.nasa.gov/20110015936.pdf

Lobo, F.S.N.; Visser, M. (25 November 2004). "Fundamental limitations on 'warp drive' spacetimes." Classical and Quantum Gravity. 21 (24): 5871. arXiv:gr-qc/0406083 doi:10.1088/0264-9381/21/24/011. "We will take the bubble velocity to be non-relativistic, $v \ll c$. Thus we are not focussing attention on the "superluminal" aspects of the warp bubble, [...] but rather on a secondary unremarked effect: The warp drive (if it can be realised in nature) appears to be an example of a "reaction-less drive" wherein the warp bubble moves by interacting with the geometry of spacetime instead of expending reaction mass." Petit, J.P.; d'Agostini, G. (10 November 2014). "Cosmological bimetric model with interacting positive and negative masses and two different speeds of light, in agreement with the observed acceleration of the Universe." Modern Physics Letters A. 29 (34). doi:10.1142/S021773231450182X.

http://www.nature.com/scientificamerican/journal/v301/n2/full/scientificamerican0809-38.html "Swimming Through Empty Space." Science 2.0.

United States gravity control propulsion research
From Wikipedia, the free encyclopedia

American interest in "gravity control propulsion research" intensified during the early 1950s. Literature from that period used the terms anti-gravity, anti-gravitation, baricentric, counterbary, electrogravitics (eGrav), G-projects, gravitics, gravity control, and gravity propulsion. Their publicized goals were to develop and discover technologies and theories for the manipulation of gravity or gravity-like fields for propulsion. Although general relativity theory appeared to prohibit anti-gravity propulsion, several programs were funded to develop it through gravitation research from 1955 to 1974. The names of many contributors to general relativity and those of the golden age of general relativity have appeared among documents about the institutions that had served as the theoretical research components of those programs. The existence and 1950s emergence of the gravity control propulsion research have not been a subject of controversy for aerospace writers, critics, and conspiracy theory advocates, but their rationale, effectiveness, and longevity have been the objects of contested views.

Evidence of existence

Mainstream newspapers, popular magazines, technical journals, and declassified papers reported the existence of the gravity control propulsion research. For example, the title of the March 1956 Aero Digest article about the intensified interest was "Anti-gravity Booming." A. V. Cleaver made the following statement about the programs in his article:

What are the facts, insofar as they are publicly known, or (as at this date) knowable? Well, they seem to amount to this: The Americans have decided to look into the old science-fictional dream of gravity control, or "anti-gravity," to investigate, both theoretically and (if possible) practically the fundamental nature of gravitational fields and their relationship to electromagnetic and other phenomena – and someone (unknown to the present writer) has apparently decided to call all this study by the high-sounding name of "electro-gravitics." Unknown, too – at least unannounced – is the name of agency or individual who decided to encourage, stimulate, or sponsor this effort, also in just what way it is being done. However, that the effort is in progress there can be little doubt, and, of course, it is entirely to be welcomed.

The gravitics programs had not been evinced by any technological artifacts, such as the Project Pluto Tory IIA, the world's first nuclear ramjet. Commemorative monuments by the Gravity Research Foundation have been the artifacts attesting to the early commitments to finding materials and methods to manipulate gravity. The endeavor had the resources and publicity of an initiative, but writers from that period did not describe them with that term. Gladych stated:

At least 14 United States universities and other research centers are hard at work cracking the gravity barrier. And backing the basic research with multi-million dollar secret projects is our aircraft industry.

The writings about the gravity control propulsion research effort had disclosed the "players" and resources while prudently withholding both the specific features of the research and the identity of its coordinating body. Publicized and telecasted conspiracy theory anecdotes have suggested much higher levels of success to the G-projects than mainstream science.

Histories

Recent historical analysis and reports have attracted attention to the agencies and firms that had participated in the gravity control propulsion research. James E. Allen, BAE Systems consultant and engineering professor at Kingston University, referred to those programs in his history of novel propulsion systems for the journal Progress in Aerospace Sciences. Research by Dr. David Kaiser, Associate Professor of the History of Science, Massachusetts Institute of Technology, manifested the contributions made by the Gravity Research Foundation to the pedagogical aspects of the golden age of general relativity.

Dr. Joshua Goldberg, Syracuse University, described the Air Force's support of relativity research during that period. Progress reports and anecdotes and Internet resumes of former visiting and staff scientists have been the sources of the history of the Research Institute for Advanced Study (RIAS). Former aviation editor of *Jane's Defense Weekly*, Nick Cook, drew attention to the antigravity programs through worldwide publications of his book, *The Hunt for Zero Point*, and subsequent televised documentaries.

Mainstream historical accounts of the G-projects have been supplemented with conspiracy theory anecdotes. Contemporaneous literature

Lists of the research institutes, industrial sites, and policy makers along with statements from prominent physicists were provided in five comprehensive works that had been published during the early years of the gravity control propulsion research. Aviation Studies (International) Limited, London, published a detailed report about those activities by the Gravity Research Group that was later declassified.

The Journal of the British Interplanetary Society and The Aeroplane published the propulsion survey and critical assessment of the American gravitics research by the internationally recognized astronautics historian A. V. Cleaver. The New York Herald Tribune and Miami Herald published a series of three articles by one of the world's greatest aviation journalists of the twentieth century, Ansel Talbert. Talbert's two series of newspaper articles took place in the midst of the policy-by-press-release era. Neither his nor the writings that followed the five prominent works from that period, yielded denials and/or retractions.

UFO and conspiracy theory literature

Gravity control propulsion research had been the subject of widely published UFO and conspiracy theory literature. The documented testimonies of whistleblowers edited by Dr. Steven Greer, Director of the Disclosure Project; anecdotes and schematics by Mark McCandlish and Milton William Cooper; and the reports by Philip J. Corso, David Darlington, and Donald Keyhoe, famous UFO researcher, have suggested incorporation of reverse engineering of recovered extraterrestrial vehicles with the anti-gravity propulsion projects had enabled them to continue beyond 1973 to successfully manufacture antigravity vehicles. Branches of the military and defense agencies have denied and refuted such claims.

Theoretical research agencies

Talbert indicated the rationale for the intensified interest in gravity control propulsion research had stemmed from the works of three physicists. They were Bryce DeWitt's prize-winning Gravity Research Foundation essay; the book *Gravity and the Universe* by Pascual Jordan; and presentations to the International Astronautical Federation by Dr. Burkhard Heim. DeWitt's essay discouraged the pursuit of materials that shield, reflect, and/or insulate gravity and emphasized the need to encourage young physicists to pursue gravitational research. He opened his essay with the following paragraph:

> Before anyone can have the audacity to formulate even the most rudimentary plan of attack on the problem of harnessing the force of gravitation, he must understand the nature of his adversary. I take it as most axiomatic that the phenomenon of gravitation is poorly understood even by the best of minds, and the last word on it is very far indeed from having been spoken.

Several articles cited his essay during and after the gravity control propulsion research period. Within a few years facilities emerged embodying the theme of

DeWitt's call for increased stimuli for research.

Physical principle surveys by Cleaver and Weyl stated the antigravity research was not based on any recognized theoretical breakthroughs. Cleaver's skepticism suggested an alternative rationale for establishing that research was based on a science fiction novel. Weyl charged publishers with poor journalism; attacked their terminology; and gave the highest rating for prospective physical principles for gravity control propulsion to Burkhard Heim's works. Stambler leveled harsh criticisms against Gluraheff's gravitation hypothesis. Talbert and other authors listed the following three agencies as the principal facilities that had conducted the theoretical research:

Gravity Research Foundation

Several articles contained expressions of gratitude for the support to the gravity control propulsion endeavor by the Gravity Research Foundation. Even though the Foundation was a humble, non-profit organization, its creator, Roger Babson, used his wealth and influence to mobilize industries; raise private and government funding; and motivate engineers and physicists to conduct research in gravity shielding and control. According to his autobiography:

"The purpose of the Foundation is to encourage others to work on gravity problems and aid others in obtaining rewards for their efforts."

During Babson's lifetime, the Foundation conducted Gravity Day Conferences each summer; established a library on gravity; solicited essays that addressed (1.) various prospects for shielding gravity, (2.) the development and/or discovery of materials that could convert gravitational force into heat, or (3.) methods of manipulating gravity; and installed monuments at various universities that cited its antigravity focus.

Aerospace Research Laboratories

In September, 1956, the General Physics Laboratory of the Aeronautical Research Laboratories (ARL) at Wright-Patterson Air Force Base, Dayton, Ohio, commenced an intense program to coordinate research into gravitational and unified field theories with the hiring of Joshua N. Goldberg. Creation by ARL of Goldberg's program may have been coincidental to the Talbert's disclosures of commitments to gravity control propulsion research. The precise rationale for creating the program and justifying its budgets and personnel may never be determined. Neither Goldberg nor the Air Force's Deputy for Scientific and Technical Information, Walter Blados, were able to locate the founding documents. Roy Kerr, a former ARL scientist, stated the antigravity propulsion purpose of ARL was "rubbish" and that "The only real use that the USAF made of us was when some crackpot sent them a proposal for antigravity or for converting rotary motion inside a spaceship to a translational driving system." The December 30, 1957 issue of Product Engineering closed its report with the following statement:

> Nevertheless, the Air Force is encouraging research in electrogravitics, and many companies and individuals are working on the problem. It could be that one of them will confound the experts.

During the following sixteen years, its name was changed to the Aerospace Research Laboratories. The ARL scientists produced nineteen technical reports and over seventy peer-reviewed journal articles. The Air Force's Foreign Technical Division, and other agencies, investigated stories about Soviet attempts to understand gravity. Such actions were consistent with the paranoia of the Cold War. The funding for the military components of the gravity control propulsion research had been terminated by the Mansfield Amendment of 1973. Black project experts, conspiracy theorists, and whistleblowers had suggested the gravity control propulsion efforts had achieved their goals and had been continued decades beyond 1973.

Research Institute for Advanced Study (RIAS)

The Research Institute for Advanced Study (RIAS) was conceived by George S. Trimble, the vice president for aviation and advanced propulsion systems, Glenn L. Martin Company, and was placed under the direct supervision of Welcome Bender. The first person Bender hired was Louis Witten internationally recognized authority on gravitation physics.

Talbert's article had announced Trimble's completion of contractual agreements with Pascual Jordan and Burkhard Heim for RIAS. Subsequent hires yielded a half dozen gravity researchers known as the field theory group. Sir Arthur C. Clarke and others stated RIAS' assemble of talent was very qualified for the task of discovering new principles that could be used to develop gravity control propulsion systems.

The quest for propulsion through gravity control was vaguely implied in various publications. Works by Cook and Cleaver summarized statements in the RIAS brochures. Cook had equated the broad range of RIAS's mission statements with those of Skunk Works. In 1958, Mallan reported "the control of the force of gravity itself for propulsion" was one of the unorthodox goals initiated by Trimble for RIAS.

RIAS was renamed the Research Institute for Advanced Studies during the sixties when the American-Marietta Company merged with Martin to become the Martin Marietta Company. The 1995 merger that yielded the Lockheed Martin Company modified its goals and not its name.

Aerospace firms

Talbert's newspaper series and subsequent articles in technical magazines and journals listed the names of aerospace firms conducting gravity control propulsion research.

The Gravity Research Group indicated those companies had constructed "rigs" to improve the performance of Thomas Townsend Brown's gravitators through attempts to develop materials with high dielectric constants (k). Gravity Rand Limited provided a set of guidelines to help management conduct research and nurture creativity.

Articles about the gravity propulsion research by the aerospace firms ceased after 1974. None of the companies featured in those publications had filed retractions. The following aerospace firms have been cited in the works published from 1955 through 1974:

Bell Aircraft, Buffalo, New York.
Boeing Aircraft.
Clarke Electronics, Palm Springs, California.
Convair, San Diego.
Douglas Aircraft.
Electronics Division, Ryan Aeronautical Company, San Diego.
General Electric.
Glenn L. Martin Company, Baltimore, Maryland.
Gluhareff Helicopter & Airplane Corporation, Manhattan Beach, California.
Grumman Aircraft.
Hiller Aircraft.
Hughes Aircraft.
Lear Incorporated, Santa Monica, California.
Lockheed Aircraft Corporation.
Radio Corporation.
Sikorsky Division of United Aircraft.
Sperry Gyroscope Division of Sperry Rand Corporation, Great Neck, Long Island.

Reported breakthroughs

None of the reported experimental breakthroughs published during the 1950s and 1960s have been recognized by the aerospace community.

Experimental Brown's gravitator

Various reports indicated Brown's gravitators were the main experimental focus of the gravity control propulsion research. According to G. Harry Stine and Intel, research on Brown's gravitators became classified immediately after demonstrations of 30% weight reductions. Thomas Townsend Brown had obtained a British patent for high voltage, symmetric, parallel plate capacitors, that he called gravitators, in 1928. Brown claimed they would produce a net thrust in the direction of the anode of the capacitor that varied slightly with the positions of the Moon. The scientific community rejected such claims as products of pseudoscience and/ or misinterpretations of ion wind effects.

Independent research found small amount of lift from Brown's gravitator based on an inefficient use of ionic propulsion. The devices were named Ion Lifters or Ionocraft and were reported to be able to lift the empty shell of a vehicle under ideal conditions, but not the additional machinery required to generate the electric field. Gravity effects were not found in the independent research.

Kaplan's gravity-like impulses

In July 1960, Missiles and Rockets reported Martin N. Kaplan, Senior

Los Angeles Times

C C TUESDAY MORNING, APRIL 8, 1952 Times

LIGHT ON MYSTERY—Watching two model flying saucers hanging from pivoting arm are trio of the new University for Social Research: Researchers Bradford Shank, left, and Townsend Brown, and Mason Rose, president. They present a novel theory.

Times photo

Flying Saucers 'Explained' by Men of New Research University Here

Two metal-plexiglass disks, suspended from a central pylon, swung through slow circles in a darkened room yesterday as a spokesman for a new university sought to convince newsmen they have solved the flying saucer mystery.

"We have hesitated to divulge our findings," said Mason Rose, president of the University for Social Research, "because they read too much like science fiction . . ."

Substance of the alleged discovery, credited to inventor Townsend Brown, is that saucers operate in a field of "electrogravity" that "acts like a wave with the negative pole at the top and the positive pole at the bottom."

Travel Like Surfboard

"The saucer travels like a surfboard on the incline of a wave that is kept continually moving by the saucer's electrogravitational generator," explained Bradford Shank, third spokesman for the group claiming knowledge "almost too sensational, too spectacular."

All three men are convinced that flying saucers are real, "controlled by an intelligence rather than a pilot" and capable of speeds up to that of light—186,000 miles a second.

Their research is new and novel, they insist, and "it is distinctly improbable it has been duplicated anywhere in the world," experiments coupling electricity and gravitation that apparently go even beyond Einstein's unified field theory.

Asked about official government study of their findings, Rose said details had been given to "some Navy admirals" but as yet there was no censorship. He talked guardedly about military "interest" in the work but declined to mention specific agencies.

He spoke too about the early trials and tribulations of Marconi, Edison and the Wright brothers.

The three men said space travel will be possible within 10 years.

At one point Shank was asked if he had a degree.

'Superior Intelligence'

"No," he acknowledged, "I'm free of those encumbrances. That's why I find it so easy to talk in these new terms."

To all dead-end questions there was the answer: "A superior intelligence thousands of years ahead of ours would have many answers we don't know about."

For more than four years Brown has been attempting to predict the ups and downs of the stock market with electronic apparatus he installed in the basement of a building on S Spring St. His equipment, he said, registers small variations in sidereal or cosmic rays which bombard the earth from outer space.

These rays, in some yet unexplained manner, are suspected of influencing human psychology. Brown declined to say how his stock market "barometer" has worked.

Rain, Holiday Snarl Traffic

It looked like the day before Christmas in the downtown area yesterday as slowly moving traffic was backed up solid from Temple to 12th Sts. and crowds crammed the sidewalks.

The really big crush was at noon.

Contributing factors: (1) rain-slickened streets, which always slows traffic; (2) pre-Easter shopping; (3) youngsters out of school for Easter vacation.

Throughout the city, of course, there was the customary epidemic of minor collisions, most of them resulting in locked bumpers—and tempers.

PHYSICIST T. TOWNSEND BROWN AND HIS ELECTROKINETIC APPARATUS

Research Engineer, Electronics Division, Ryan Aeronautical Company, San Diego, had conducted anti-gravitational experiments yielding the promise of impulses, accelerations, and decelerations one hundred times the pull of gravity. Neither comments nor criticism of the report appeared in subsequent articles during the period of intensified gravity control propulsion research (see Section 1 of tractor beam for similar reports).

Theoretical
Forward's protational field

Robert L. Forward, Hughes Research Laboratories, Malibu, described the theoretical generation of dipole gravitational fields by accelerating a super-dense fluid through pipes wound around a torus.

Legacies

Many of the contributors to general relativity have been supported by and/ or associated with the ARL, RIAS, and/or the Gravity Research Foundation. The decades preceding the 1955 revelation of the gravity control propulsion research were a low water mark for general relativity.

The following summarizes how the components of that research had stimulated the resurgence of general relativity: Gravity Research Foundation Even though some of the physicists who attended the Gravity Day Conferences quietly mocked the anti-gravity mission of the Foundation, it provided significant contributions to mainstream physics. The International Journal of Modern Physics D has featured selected papers from the Gravity Research Foundation essay competition. Many have been incorporated with the collections of the Niels Bohr Library. A few of the Foundation essay contest winners became Nobel laureates (e.g., Ilya Prigogine, Maurice Allais, George F. Smoot). Foundation essays have been among the resources graduate students check for new ideas. Kaiser summarized the Foundation's influence in the following manner:

> Despite the vast conceptual gulf separating Babson from the new generation of relativists, we are left with intriguing, and perhaps ironic associations: by organizing conferences, sponsoring the annual essay contests, and making money and enthusiasm widely available for people interested in gravity, the eccentric Gravity Research Foundation may claim at least some small amount of the credit for helping to stimulate the postwar resurgence of interest in gravitation and general relativity.

Foundation trustee, Agnew Bahnson, contacted Dr. Bryce DeWitt with a proposal to fund the creation of a gravity research institute. DeWitt had won the first prize for the 1953 essay contest. The proposed name was changed to the Institute for Field Physics and it was established in 1956 at the University of North Carolina at Chapel Hill under the direction of Bryce and his wife, Cécile DeWitt-Morette.

The peer reviewed physics journal, Physica C, published a report by Eugene Podkletnov and Nieminen about gravity-like shielding. Although their work had gained international attention, researchers were not able to replicate Podkletnov's initial conditions. But, analyses by Giovanni Modanese and Ning Wu indicated various applications of quantum gravity theory could allow gravitational shielding phenomena. Those achievements have not been pursued by the scientific community.

Aerospace Research Laboratories (ARL)

The list of prominent contributors to the golden age of general relativity, contains the names of several scientists who had authored the nineteen ARL Technical Reports and/or seventy papers. The ARL sponsored papers were published in the Proceedings of the Royal Society of London, Physical Review, Journal of Mathematical Physics, Physical Review Letters, Physical Review D, Review of Modern Physics, General Relativity and Gravitation, International Journal of Theoretical Physics, and Nuovo Cimento B. Some of the ARL papers were written in collaboration with RIAS, the U.S. Army Signal Research and Development Laboratory at Fort Monmouth, New Jersey, and the Office of Naval Research. The ARL had provided significant enhancements to general relativity theory. For example, Roy Kerr's description of the behavior of space-time in the vicinity of a rotating mass was among those works.[55] Goldberg concluded: "However, it should be recognized that, in the United States, the Department of Defense played an essential role in building a strong scientific community without widespread encroachment on academic values."

Research Institute for Advanced Studies (RIAS)

The growth of nonlinear differential equations during the fifties was stimulated by RIAS. One of the leading groups in dynamical systems and control theory, the Lefschetz Center for Dynamical Systems, was a spinoff from RIAS. After the launch of Sputnik, world-class mathematician Solomon Lefschetz came out of retirement to join RIAS in 1958 and formed the world's largest group of mathematicians devoted to research in nonlinear differential equations. The RIAS mathematics group stimulated the growth of nonlinear differential equations through conferences and publications. It left RIAS in 1964 to form the Lefschetz Center for Dynamical Systems at Brown University, Providence, Rhode Island.

UFO and conspiracy theories

On May 9, 2001, Mark McCandlish testified on the televised news conference held by the Disclosure Project, at the National Press Club, Washington, D.C. He stated gravity control propulsion research had started in the fifties and had successfully reverse engineered the vehicle retrieved from the Roswell crash site to build three Alien Reproduction Vehicles by 1981. McCandlish described their propulsion systems in terms of Thomas Townsend Brown's gravitators and provided a line drawing of its interior. The diagram closely resembled the drawing provided earlier in Milton William Cooper's book. Another Disclosure Project whistleblower, Philip J. Corso, stated in his book the craft retrieved from the second crash site at Roswell, New Mexico, had a propulsion system resembling Thomas Townsend Brown's gravitators. And, Corso's book

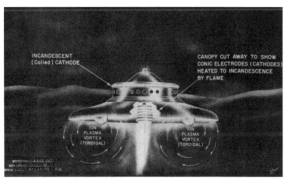

featured several gravity control propulsion statements made by Hermann Oberth.

Soon after the end of the Cold War, a small group of scientists and engineers openly expressed their desire to use technologies developed by black projects for civil applications. Steven Greer formed the Disclosure Project in 1995 to help those and other research whistleblowers share their information with and to petition Congress. By 2001, it had provided reports to two Congressional hearings and had acquired over 400 members from branches of the military and aerospace industry.

During the early 1960s, Keyhoe published excerpts from a letter by Hermann Oberth that presented explanations for the flight characteristics of UFO's in terms of gravity control propulsion. Prior to Oberth's letter, Keyhoe had supported arguments for magnetic forces as the source of propulsion for UFO's. The letter caused him to search for the existence of gravity control propulsion research programs. The following is a segment of his findings he had released in his 1966 and 1974 publications:

> When AF researchers fully realized the astounding possibilities, headquarters persuaded scientists, aerospace companies and technical laboratories to set up anti-gravity projects, many of them under secret contracts. Every year, the number of projects increased. In 1965, forty-six unclassified G-projects were confirmed to me by the Scientific Information Exchange of the Smithsonian Institution. Of the forty-six, thirty-three were AF-controlled.

During his press conferences on February 2, 1955 in Bogotá, and February 10, 1955 in Grand Rapids, Michigan aviation pioneer William Lear, stated one of his reasons for believing in flying saucers was the existence of American research efforts into antigravity. Talbert's series of newspaper articles about the intensified interest in gravity control propulsion research were published during the Thanksgiving week of that year.

ELECTROKINETIC GRAVITATORS
TT BROWN

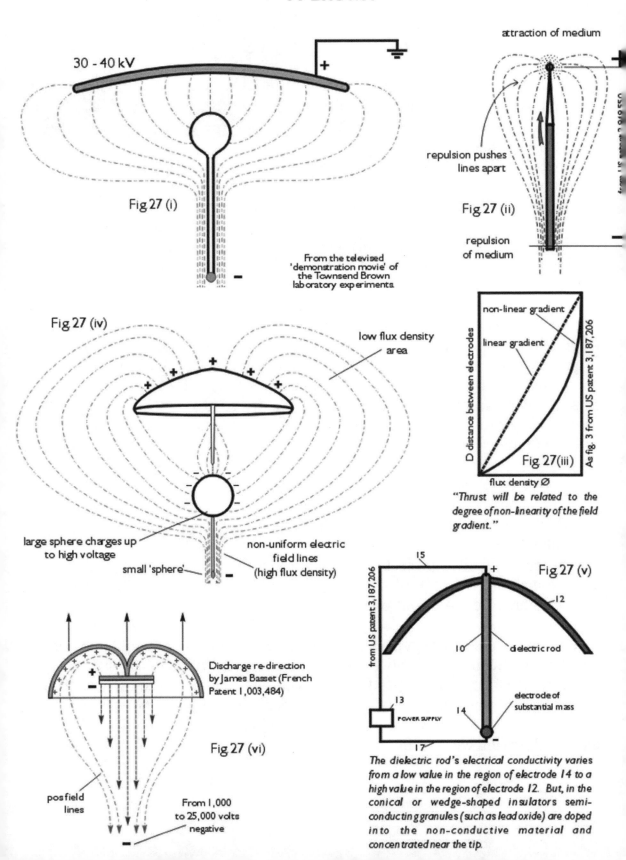

30 - 40 kV

+

Fig 27 (i)

From the televised 'demonstration movie' of the Townsend Brown laboratory experiments.

attraction of medium

repulsion pushes lines apart

repulsion of medium

Fig 27 (ii)

from US patent 3,819,850

Fig 27 (iv)

low flux density area

+ + + +

large sphere charges up to high voltage

small 'sphere'

non-uniform electric field lines (high flux density)

non-linear gradient

linear gradient

D distance between electrodes

flux density Ø

Fig 27(iii)

As fig. 3 from US patent 3,187,206

"Thrust will be related to the degree of non-linearity of the field gradient."

15

+

Fig 27 (v)

12

from US patent 3,187,206

10   dielectric rod

13

POWER SUPPLY

14

electrode of substantial mass

17

The dielectric rod's electrical conductivity varies from a low value in the region of electrode 14 to a high value in the region of electrode 12. But, in the conical or wedge-shaped insulators semi-conducting granules (such as lead oxide) are doped into the non-conductive material and concentrated near the tip.

+
−

Discharge re-direction by James Basset (French Patent 1,003,484)

Fig 27 (vi)

pos field lines

From 1,000 to 25,000 volts negative

Chapter 2

The Motionless Electromagnet Generator Patent

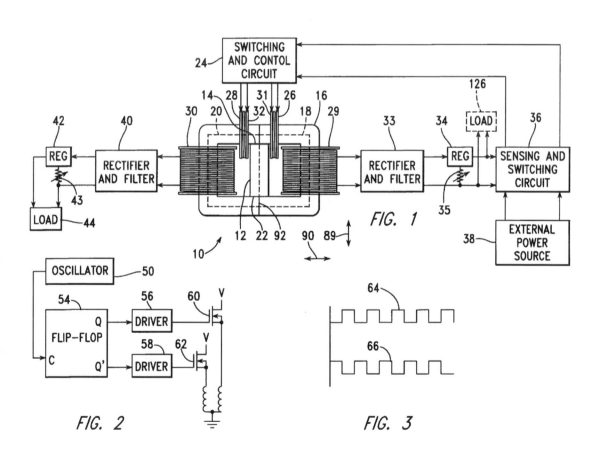

FIG. 1

FIG. 2

FIG. 3

US006362718B1

(12) **United States Patent**
Patrick et al.

(10) **Patent No.:** **US 6,362,718 B1**
(45) **Date of Patent:** **Mar. 26, 2002**

(54) **MOTIONLESS ELECTROMAGNETIC GENERATOR**

(76) Inventors: **Stephen L. Patrick**, 2511 Woodview Dr. SE.; **Thomas E. Bearden**, 2211 Cove Rd., both of Huntsville, AL (US) 35801; **James C. Hayes**, 16026 Deaton Dr. SE., Huntsville, AL (US) 35803; **Kenneth D. Moore**, 1704 Montdale Rd., Huntsville, FL (US) 35801; **James L. Kenny**, 925 Tascosa Dr., Huntsville, AL (US) 35802

(*) Notice: Subject to any disclaimer, the term of this patent is extended or adjusted under 35 U.S.C. 154(b) by 0 days.

(21) Appl. No.: **09/656,313**

(22) Filed: **Sep. 6, 2000**

(51) Int. Cl.[7] H01F 27/24
(52) U.S. Cl. **336/214**
(58) Field of Search 363/16, 24, 25, 363/26, 56.06, 56.08, 133, 134; 336/15, 110, 155, 177, 180, 213, 214, 221, 222

(56) **References Cited**

U.S. PATENT DOCUMENTS

2,153,378 A	4/1939	Kramer	171/95
2,892,155 A	6/1959	Radus et al.	324/117
3,079,535 A	2/1963	Schultz	317/201
3,165,723 A	1/1965	Radus	340/174
3,228,013 A	1/1966	Olson et al.	340/174
3,254,268 A	5/1966	Radus et al.	317/14
3,316,514 A	4/1967	Radus et al.	335/291
3,368,141 A	2/1968	Subieta-Garron	323/44
3,391,358 A	7/1968	Bratkowski et al.	335/21
3,453,876 A	7/1969	Radus	73/141
3,517,300 A	* 6/1970	McMurray	
3,569,947 A	3/1971	Radus	340/174
3,599,074 A	* 8/1971	Adams	
4,006,401 A	2/1977	de Rivas	323/92
4,077,001 A	2/1978	Richardson	323/92
4,366,532 A	12/1982	Rosa et al.	363/69
4,482,945 A	11/1984	Wolf et al.	363/129
4,554,524 A	11/1985	Radus	337/3

4,853,668 A	8/1989	Bloom	336/214
4,864,478 A	9/1989	Bloom	363/16
4,904,926 A	2/1990	Pasichinskyj	323/362
5,011,821 A	4/1991	McCullough	505/1
5,221,892 A	6/1993	Sullivan et al.	323/362
5,245,521 A	* 9/1993	Spreen	363/37
5,327,015 A	7/1994	Hacket	505/211
5,335,163 A	8/1994	Seiersen	363/126
5,694,030 A	* 12/1997	Sato et al.	323/282

OTHER PUBLICATIONS

Raymond J. Radus, "Permanent–Magnet Circuit using a 'Flux–Transfer' Principle," Engineers' Digest, 24(1–6) Jan.–Jun. 1963, p. 86.
Robert O'Handley, Modern Magnetic Materials, Principles and Applications, John Wiley & Sons, Inc., 2000, pp. 456–468.
Robert C. Weast, Editor, CRC Handbook of Chemistry and Physics, 1978–1979, p. B–50.
Honeywell.com web site, "amorphous metals".

* cited by examiner

Primary Examiner—Matthew Nguyen
(74) *Attorney, Agent, or Firm*—Norman Friedland

(57) **ABSTRACT**

An electromagnetic generator without moving parts includes a permanent magnet and a magnetic core including first and second magnetic paths. A first input coil and a first output coil extend around portions of the first magnetic path, while a second input coil and a second output coil extend around portions of the second magnetic path. The input coils are alternatively pulsed to provide induced current pulses in the output coils. Driving electrical current through each of the input coils reduces a level of flux from the permanent magnet within the magnet path around which the input coil extends. In an alternative embodiment of an electromagnetic generator, the magnetic core includes annular spaced-apart plates, with posts and permanent magnets extending in an alternating fashion between the plates. An output coil extends around each of these posts. Input coils extending around portions of the plates are pulsed to cause the induction of current within the output coils.

29 Claims, 5 Drawing Sheets

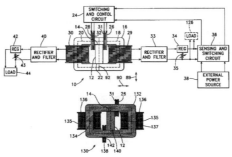

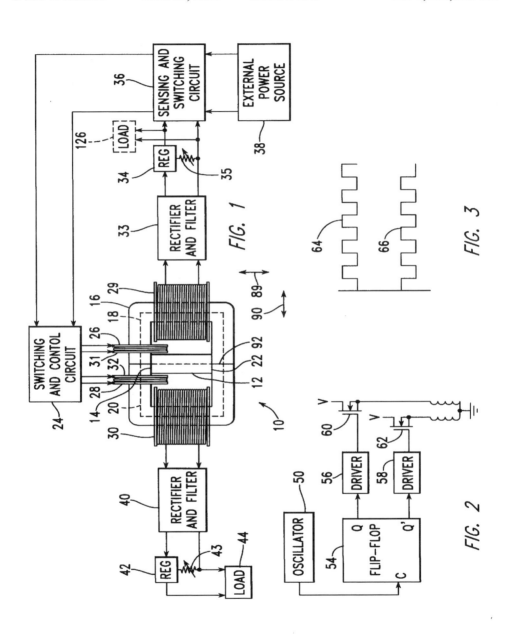

FIG. 1

FIG. 2

FIG. 3

U.S. Patent   Mar. 26, 2002   Sheet 2 of 5   **US 6,362,718 B1**

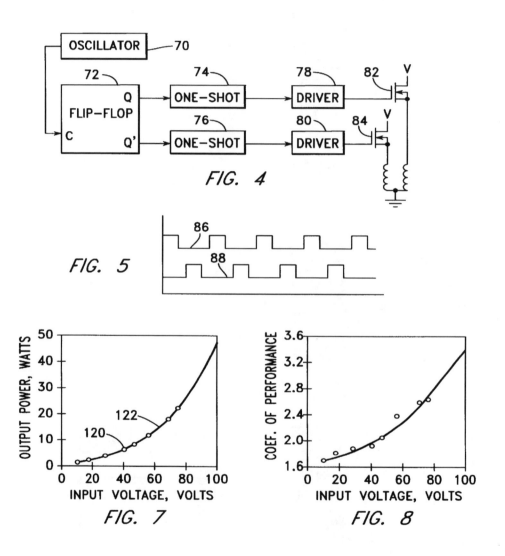

FIG. 4

FIG. 5

FIG. 7

FIG. 8

U.S. Patent   Mar. 26, 2002   Sheet 3 of 5   US 6,362,718 B1

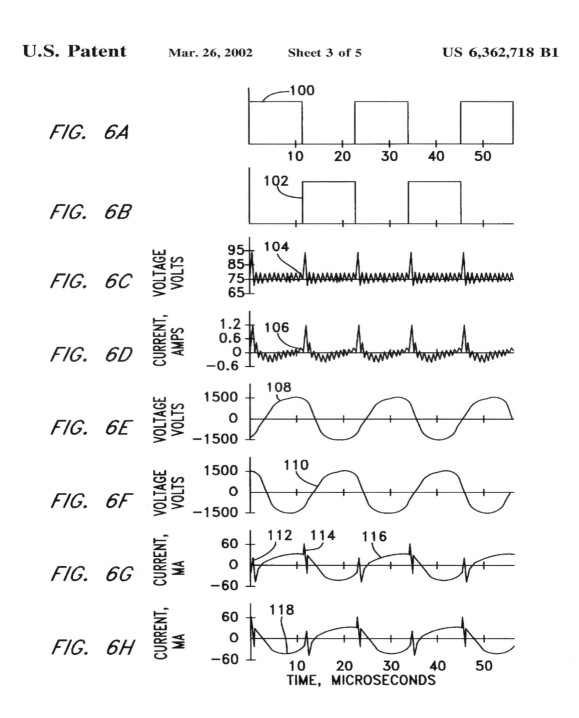

FIG. 6A

FIG. 6B

FIG. 6C

FIG. 6D

FIG. 6E

FIG. 6F

FIG. 6G

FIG. 6H

The Anti-Gravity Files

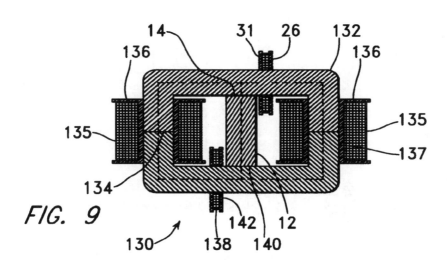

FIG. 9

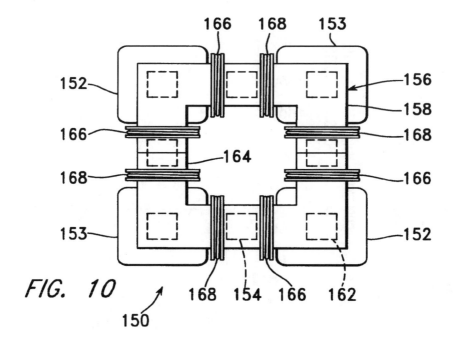

FIG. 10

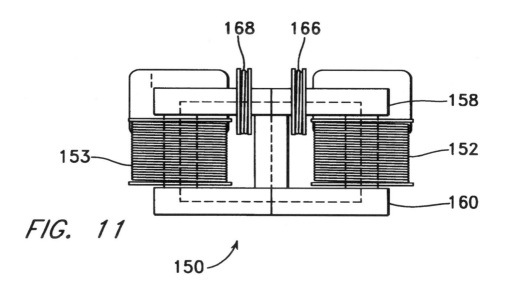

FIG. 11

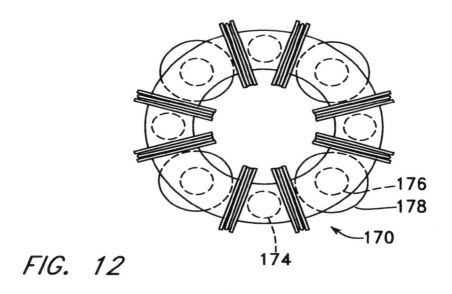

FIG. 12

US 6,362,718 B1

1

MOTIONLESS ELECTROMAGNETIC GENERATOR

BACKGROUND INFORMATION

1. Field of Invention

This invention relates to a magnetic generator used to produce electrical power without moving parts, and, more particularly, to such a device having a capability, when operating, of producing electrical power without an external application of input power through input coils.

2. Description of the Related Art

The patent literature describes a number of magnetic generators, each of which includes a permanent magnet, two magnetic paths external to the permanent magnet, each of which extends between the opposite poles of the permanent magnet, switching means for causing magnetic flux to flow alternately along each of the two magnetic paths, and one or more output coils in which current is induced to flow by means of changes in the magnetic field within the device. These devices operate in accordance with an extension of Faraday's Law, indicating that an electrical current is induced within a conductor within a changing magnetic field, even if the source of the magnetic field is stationary.

A method for switching magnetic flux to flow predominantly along either of two magnetic paths between opposite poles of a permanent magnet is described as a "flux transfer" principle by R. J. Radus in *Engineer's Digest*, Jul. 23, 1963. This principle is used to exert a powerful magnetic force at one end of both the north and south poles and a very low force at the other end, without being used in the construction of a magnetic generator. This effect can be caused mechanically, by keeper movement, or electrically, by driving electrical current through one or more control windings extending around elongated versions of the pole pieces **14**. Several devices using this effect are described in U.S. Pat. Nos. 3,165,723, 3,228,013, and 3,316,514, which are incorporated herein by reference.

Another step toward the development of a magnetic generator is described in U.S. Pat. No. 3,368,141, which is incorporated herein by reference, as a device including a permanent magnet in combination with a transformer having first and second windings about a core, with two paths for magnetic flux leading from each pole of the permanent magnet to either end of the core, so that, when an alternating current induces magnetic flux direction changes in the core, the magnetic flux from the permanent magnet is automatically directed through the path which corresponds with the direction taken by the magnetic flux through the core due to the current. In this way, the magnetic flux is intensified. This device can be used to improve the power factor of a typically inductively loaded alternating current circuit.

Other patents describe magnetic generators in which electrical current from one or more output coils is described as being made available to drive a load, in the more conventional manner of a generator. For example, U.S. Pat. No. 4,006,401, which is incorporated herein by reference, describes an electromagnetic generator including permanent magnet and a core member, in which the magnetic flux flowing from the magnet in the core member is rapidly alternated by switching to generate an alternating current in a winding on the core member. The device includes a permanent magnet and two separate magnetic flux circuit paths between the north and south poles of the magnet. Each of the circuit paths includes two switching means for alternately opening and closing the circuit paths, generating an alternating current in a winding on the core member. Each

2

of the switching means includes a switching magnetic circuit intersecting the circuit path, with the switching magnetic circuit having a coil through which current is driven to induce magnetic flux to saturate the circuit path extending to the permanent magnet. Power to drive these coils is derived directly from the output of a continuously applied alternating current source. What is needed is an electromagnetic generator not requiring the application of such a current source.

U.S. Pat. No. 4,077,001, which is incorporated herein by reference, describes a magnetic generator, or dc/dc converter, comprising a permanent magnet having spaced-apart poles and a permanent magnetic field extending between the poles of the magnet. A variable-reluctance core is disposed in the field in fixed relation to the magnet and the reluctance of the core is varied to cause the pattern of lines of force of the magnetic field to shift. An output conductor is disposed in the field in fixed relation to the magnet and is positioned to be cut by the shifting lines of permanent magnetic force so that a voltage is induced in the conductor. The magnetic flux is switched between alternate paths by means of switching coils extending around portions of the core, with the flow of current being alternated between these switching coils by means of a pair of transistors driven by the outputs of a flip-flop. The input to the flip flop is driven by an adjustable frequency oscillator. Power for this drive circuit is supplied through an additional, separate power source. What is needed is a magnetic generator not requiring the application of such a power source.

U.S. Pat. No. 4,904,926, which is incorporated herein by reference, describes another magnetic generator using the motion of a magnetic field. The device includes an electrical winding defining a magnetically conductive zone having bases at each end, the winding including elements for the removing of an induced current therefrom. The generator further includes two pole magnets, each having a first and a second pole, each first pole in magnetic communication with one base of the magnetically conductive zone. The generator further includes a third pole magnet, the third pole magnet oriented intermediately of the first poles of the two pole electromagnets, the third pole magnet having a magnetic axis substantially transverse to an axis of the magnetically conductive zone, the third magnet having a pole nearest to the conductive zone and in magnetic attractive relationship to the first poles of the two pole electromagnets, in which the first poles thereof are like poles. Also included in the generator are elements, in the form of windings, for cyclically reversing the magnetic polarities of the electromagnets. These reversing means, through a cyclical change in the magnetic polarities of the electromagnets, cause the magnetic flux lines associated with the magnetic attractive relationship between the first poles of the electromagnets and the nearest pole of the third magnet to correspondingly reverse, causing a wiping effect across the magnetically conductive zone, as lines of magnetic flux swing between respective first poles of the two electromagnets, thereby inducing electron movement within the output windings and thus generating a flow of current within the output windings.

U.S. Pat. No. 5,221,892, which is incorporated herein by reference, describes a magnetic generator in the form of a direct current flux compression transformer including a magnetic envelope having poles defining a magnetic axis and characterized by a pattern of magnetic flux lines in polar symmetry about the axis. The magnetic flux lines are spatially displaced relative to the magnetic envelope using control elements which are mechanically stationary relative to the core. Further provided are inductive elements which

US 6,362,718 B1

3

are also mechanically stationary relative to the magnetic envelope. Spatial displacement of the flux relative to the inductive elements causes a flow of electrical current. Further provided are magnetic flux valves which provide for the varying of the magnetic reluctance to create a time domain pattern of respectively enhanced and decreased magnetic reluctance across the magnetic valves, and, thereby, across the inductive elements.

Other patents describe devices using superconductive elements to cause movement of the magnetic flux. These devices operate in accordance with the Meissner effect, which describes the expulsion of magnetic flux from the interior of a superconducting structure as the structure undergoes the transition to a superconducting phase. For example, U.S. Pat. No. 5,011,821, which is incorporated herein by reference, describes an electric power generating device including a bundle of conductors which are placed in a magnetic field generated by north and south pole pieces of a permanent magnet. The magnetic field is shifted back and forth through the bundle of conductors by a pair of thin films of superconductive material. One of the thin films is placed in the superconducting state while the other thin film is in a non-superconducting state. As the states are cyclically reversed between the two films, the magnetic field is deflected back and forth through the bundle of conductors.

U.S. Pat. No. 5,327,015, which is incorporated herein by reference, describes an apparatus for producing an electrical impulse comprising a tube made of superconducting material, a source of magnetic flux mounted about one end of the tube, a means, such as a coil, for intercepting the flux mounted along the tube, and a means for changing the temperature of the superconductor mounted about the tube. As the tube is progressively made superconducting, the magnetic field is trapped within the tube, creating an electrical impulse in the means for intercepting. A reversal of the superconducting state produces a second pulse.

None of the patented devices described above use a portion of the electrical power generated within the device to power the reversing means used to change the path of magnetic flux. Thus, like conventional rotary generators, these devices require a steady input of power, which may be in the form of electrical power driving the reversing means of one of these magnetic generators or the torque driving the rotor of a conventional rotary generator. Yet, the essential function of the magnetic portion of an electrical generator is simply to switch magnetic fields in accordance with precise timing. In most conventional applications of magnetic generators, the voltage is switched across coils, creating magnetic fields in the coils which are used to override the fields of permanent magnets, so that a substantial amount of power must be furnished to the generator to power the switching means, reducing the efficiency of the generator.

Recent advances in magnetic material, which have particularly been described by Robert C. O'Handley in *Modern Magnetic Materials, Principles and Applications*, John Wiley & Sons, New York, pp. 456–468, provide nanocrystalline magnetic alloys, which are particularly well suited forth rapid switching of magnetic flux. These alloys are primarily composed of crystalline grains, or crystallites, each of which has at least one dimension of a few nanometers. Nanocrystalline materials may be made by heat-treating amorphous alloys which form precursors for the nanocrystalline materials, to which insoluble elements, such as copper, are added to promote massive nucleation, and to which stable, refractory alloying materials, such as niobium or tantalum carbide are added to inhibit grain growth. Most of the volume of nanocrystalline alloys is composed of

4

randomly distributed crystallites having dimensions of about 2–40 nm. These crystallites are nucleated and grown from an amorphous phase, with insoluble elements being rejected during the process of crystallite growth. In magnetic terms, each crystallite is a single-domain particle. The remaining volume of nanocrystalline alloys is made up of an amorphous phase in the form of grain boundaries having a thickness of about 1 nm.

Magnetic materials having particularly useful properties are formed from an amorphous Co—Nb—B (cobalt-niobium-boron) alloy having near-zero magnetostriction and relatively strong magnetization, as well as good mechanical strength and corrosion resistance. A process of annealing this material can be varied to change the size of crystallites formed in the material, with a resulting strong effect on DC coercivity. The precipitation of nanocrystallites also enhances AC performance of the otherwise amorphous alloys.

Other magnetic materials are formed using iron-rich amorphous and nanocrystalline alloys, which generally show larger magnetization. Such materials are, for example, Fe—B—Si—Nb—Cu (iron-boron-silicon-niobium-copper) alloys. While the permeability of iron-rich amorphous alloys is limited by their relatively large levels of magnetostriction, the formation of a nanocrystalline material from such an amorphous alloy dramatically reduces this level of magnetostriction, favoring easy magnetization.

Advances have also been made in the development of materials for permanent magnets, particularly in the development of materials including rare earth elements. Such materials include samarium cobalt, $SmCo_5$, which is used to form a permanent magnet material having the highest resistance to demagnetization of any known material. Other magnetic materials are made, for example, using combinations of iron, neodymium, and boron.

SUMMARY OF THE INVENTION

It is a first objective of the present invention to provide a magnetic generator which a need for an external power source during operation of the generator is eliminated.

It is a second objective of the present invention to provide a magnetic generator in which a magnetic flux path is changed without a need to overpower a magnetic field to change its direction.

It is a third objective of the present invention to provide a magnetic generator in which the generation of electricity is accomplished without moving parts.

In the apparatus of the present invention, the path of the magnetic flux from a permanent magnet is switched in a manner not requiring the overpowering of the magnetic fields. Furthermore, a process of self-initiated iterative switching is used to switch the magnetic flux from the permanent magnet between alternate magnetic paths within the apparatus, with the power to operate the iterative switching being provided through a control circuit consisting of components known to use low levels of power. With self-switching, a need for an external power source during operation of the generator is eliminated, with a separate power source, such as a battery, being used only for a very short time during start-up of the generator.

According to a first aspect of the present invention, an electromagnetic generator is provided, including a permanent magnet, a magnetic core, first and second input coils, first and second output coils, and a switching circuit. The permanent magnet has magnetic poles at opposite ends. The

US 6,362,718 B1

5

magnetic core includes a first magnetic path, around which the first input and output coils extend, and a second magnetic path, around which the second input and output coils extend, between opposite ends of the permanent magnet. The switching circuit drives electrical current alternately through the first and second input coils. The electrical current driven through the first input oil causes the first input coil to produce a magnetic field opposing a concentration of magnetic flux from the permanent magnet within the first magnetic path. The electrical current driven through the second input coil causes the second input coil to produce a magnetic field opposing a concentration of magnetic flux from the permanent magnet within the second magnetic path.

According to another aspect of the present invention, an electromagnetic generator is provided, including a magnetic core, a plurality of permanent magnets, first and second pluralities of input coils, a plurality of output coils, and a switching circuit. The magnetic core includes a pair of spaced-apart plates, each of which has a central aperture, and first and second pluralities of posts extending between the spaced-apart plates. The permanent magnets each extend between the pair of spaced apart plates. Each permanent magnet has magnetic poles at opposite ends, with the magnetic fields of all the permanent magnets being aligned to extend in a common direction. Each input coil extends around a portion of a plate within the spaced-apart plates, between a post and a permanent magnet. An output coil extends around each post. The switching circuit drives electrical current alternately through the first and second pluralities of input coils. Electrical current driven through each input coil in the first plurality of input coils causes an increase in magnetic flux within each post within the first plurality of posts from permanent magnets on each side of the post and a decrease in magnetic flux within each post within the second plurality of posts from permanent magnets on each side of the post. Electrical current driven through each input coil in the second plurality of input coils causes a decrease in magnetic flux within each post within the first plurality of posts from permanent magnets on each side of the post and an increase in magnetic flux within each post within the second plurality of posts from permanent magnets on each side of the post.

BRIEF DESCRIPTION OF THE DRAWINGS

FIG. 1 is a partly schematic front elevation of a magnetic generator and associated electrical circuits built in accordance with a first version of the first embodiment of the present invention;

FIG. 2 is a schematic view of a first version of a switching and control circuit within the associated electrical circuits of FIG. 1;

FIG. 3 is a graphical view of drive signals produced within the circuit of FIG. 2;

FIG. 4 is a schematic view of a second version of a switching and control circuit within the associated electrical circuits of FIG. 1;

FIG. 5 is a graphical view of drive signals produced within the circuit of FIG. 3;

FIG. 6A is a graphical view of a first drive signal within the apparatus of FIG. 1;

FIG. 6B is a graphical view of a second drive signal within the apparatus of FIG. 1;

FIG. 6C is a graphical view of an input voltage signal within the apparatus of FIG. 1;

FIG. 6D is a graphical view of an input current signal within the apparatus of FIG. 1;

6

FIG. 6E is a graphical view of a first output voltage signal within the apparatus of FIG. 1;

FIG. 6F is a graphical view of a second output voltage signal within the apparatus of FIG. 1;

FIG. 6G is a graphical view of a first output current signal within the apparatus of FIG. 1;

FIG. 6H is a graphical view of a second output current signal within the apparatus of FIG. 1;

FIG. 7 is a graphical view of output power measured within the apparatus of FIG. 1, as a function of input voltage;

FIG. 8 is a graphical view of a coefficient of performance, calculated from measurements within the apparatus of FIG. 1, as a function of input voltage;

FIG. 9 is a cross-sectional elevation of a second version of the first embodiment of the present invention;

FIG. 10 is a top view of a magnetic generator built in accordance with a first version of a second embodiment of the present invention;

FIG. 11 is a front elevation of the magnetic generator of FIG. 10; and

FIG. 12 is a top view of a magnetic generator built in accordance with a second version of the second embodiment of the present invention.

DETAILED DESCRIPTION OF THE INVENTION

FIG. 1 is a partly schematic front elevation of an electromagnetic generator 10, built in accordance with a first embodiment of the present invention to include a permanent magnet 12 to supply input lines of magnetic flux moving from the north pole 14 of the magnet 12 outward into magnetic flux path core material 16. The flux path core material 16 is configured to form a right magnetic path 18 and a left magnetic path 20, both of which extend externally between the north pole 14 and the south pole 22 of the magnet 12. The electromagnetic generator 10 is driven by means of a switching and control circuit 24, which alternately drives electrical current through a right input coil 26 and a left input coil 28. These input coils 26, 28 each extend around a portion of the core material 16, with the right input coil 26 surrounding a portion of the right magnetic path 18 and with the left input coil 28 surrounding a portion of the left magnetic path 20. A right output coil 29 also surrounds a portion of the right magnetic path 18, while a left output coil 30 surrounds a portion of the left magnetic path 20.

In accordance with a preferred version of the present invention, the switching and control circuit 24 and the input coils 26, 28 are arranged so that, when the right input coil 26 is energized, a north magnetic pole is present at its left end 31, the end closest to the north pole 14 of the permanent magnet 12, and so that, when the left input coil 28 is energized, a north magnetic pole is present at its right end 32, which is also the end closest to the north pole 14 of the permanent magnet 12. Thus, when the right input coil 26 is magnetized, magnetic flux from the permanent magnet 12 is repelled from extending through the right input coil 26. Similarly, when the left input coil 28 is magnetized, magnetic flux from the permanent magnet 12 is repelled from extending through the left input coil 28.

Thus, it is seen that driving electrical current through the right input coil 26 opposes a concentration of flux from the permanent magnet 12 within the right magnetic path 18, causing at least some of this flux to be transferred to the left magnetic path 20. On the other hand, driving electrical current through the left input coil 28 opposes a concentration

US 6,362,718 B1

7

of flux from the permanent magnet **12** within the left magnetic path **20**, causing at least some of this flux to be transferred to the right magnetic path **18**.

While in the example of FIG. **1**, the input coils **26, 28** are placed on either side of the north pole of the permanent magnet **12**, being arranged along a portion of the core **16** extending from the north pole of the permanent magnet **12**, it is understood that the input coils **26, 28** could as easily be alternately placed on either side of the south pole of the permanent magnet **12**, being arranged along a portion of the core **16** extending from the south pole of the permanent magnet **12**, with the input coils **26, 28** being wired to form, when energized, magnetic fields having south poles directed toward the south pole of the permanent magnet **12**. In general, the input coils **26, 28** are arranged along the magnetic core on either side of an end of the permanent magnet forming a first pole, such as a north pole, with the input coils being arranged to produce magnetic fields of the polarity of the first pole directed toward the first pole of the permanent magnet.

Further in accordance with a preferred version of the present invention, the input coils **26, 28** are never driven with so much current that the core material **16** becomes saturated. Driving the core material **16** to saturation means that subsequent increases in input current can occur without effecting corresponding changes in magnetic flux, and therefore that input power can be wasted. In this way, the apparatus of the present invention is provided with an advantage in terms of the efficient use of input power over the apparatus of U.S. Pat. No. 4,000,401, in which a portion both ends of each magnetic path is driven to saturation to block flux flow. In the electromagnetic generator **10**, the switching of current flow within the input coils **26, 28** does not need to be sufficient to stop the flow of flux in one of the magnetic paths **18, 20** while promoting the flow of magnetic flux in the other magnetic path. The electromagnetic generator **10** works by changing the flux pattern; it does not need to be completely switched from one side to another.

Experiments have determined that this configuration is superior, in terms of the efficiency of using power within the input coils **26, 28** to generate electrical power within the output coils **29, 30**, to the alternative of arranging input coils and the circuits driving them so that flux from the permanent magnet is driven through the input coils as they are energized. This arrangement of the present invention provides a significant advantage over the prior-art methods shown, for example, in U.S. Pat. No. 4,077,001, in which the magnetic flux is driven through the energized coils.

The configuration of the present invention also has an advantage over the prior-art configurations of U.S. Pat. Nos. 3,368,141 and 4,077,001 in that the magnetic flux is switched between two alternate magnetic paths **18, 20** with only a single input coil **26, 28** surrounding each of the alternate magnetic paths. The configurations of U.S. Pat. Nos. 3,368,141 and 4,077,001 each require two input coils on each of the magnetic paths. This advantage of the present invention is significant both in the simplification of hardware and in increasing the efficiency of power conversion.

The right output coil **29** is electrically connected to a rectifier and filter **33**, having an output driven through a regulator **34**, which provides an output voltage adjustable through the use of a potentiometer **35**. The output of the linear regulator **34** is in turn provided as an input to a sensing and switching circuit **36**. Under start up conditions, the sensing and switching circuit **36** connects the switching and control circuit **24** to an external power source **38**, which is,

8

for example, a starting battery. After the electromagnetic generator **10** is properly started, the sensing and switching circuit **36** senses that the voltage available from regulator **34** has reached a predetermined level, so that the power input to the switching and control circuit **24** is switched from the external power source **38** to the output of regulator **34**. After this switching occurs, the electromagnetic generator **10** continues to operate without an application of external power.

The left output coil **30** is electrically connected to a rectifier and filter **40**, the output of which is connected to a regulator **42**, the output voltage of which is adjusted by means of a potentiometer **43**. The output of the regulator **42** is in turn connected to an external load **44**.

FIG. **2** is a schematic view of a first version of the switching and control circuit **24**. An oscillator **50** drives the clock input of a flip-flop **54**, with the Q and Q' outputs of the flip-flop **54** being connected through driver circuits **56, 58** to power FETS **60, 62** so that the input coils **26, 28** are alternately driven. In accordance with a preferred version of the present invention, the voltage V applied to the coils **26, 28** through the FETS **60, 62** is derived from the output of the sensing and switching circuit **36**.

FIG. **3** is a graphical view of the signals driving the gates of FETS **60, 62** of FIG. **2**, with the voltage of the signal driving the gate of FET **60** being represented by line **64**, and with the voltage of the signal driving FET **62** being represented by line **66**. Both of the coils **26, 28** are driven with positive voltages.

FIG. **4** is a schematic view of a second version of the switching and control circuit **24**. In this version, an oscillator **70** drives the clock input of a flip-flop **72**, with the Q and Q' outputs of the flip-flop **72** being connected to serve as triggers for one-shots **74, 76**. The outputs of the one-shots **74, 76** are in turn connected through driver circuits **78, 80** to drive FETS **82, 84**, so that the input coils **26, 28** are alternately driven with pulses shorter in duration than the Q and Q' outputs of the flip flop **72**.

FIG. **5** is a graphical view of the signals driving the gates of FETS **82, 84** of FIG. **4**, with the voltage of the signal driving the gate of FET **82** being represented by line **86**, and with the voltage of the signal driving the gate of FET **84** being represented by line **88**.

Referring again to FIG. **1**, power is generated in the right output coil **29** only when the level of magnetic flux is changing in the right magnetic path **18**, and in the left output coil **30** only when the level of magnetic flux is changing in the left magnetic path **20**. It is therefore desirable to determine, for a specific magnetic generator configuration, the width of a pulse providing the most rapid practical change in magnetic flux, and then to provide this pulse width either by varying the frequency of the oscillator **50** of the apparatus of FIG. **2**, so that this pulse width is provided with the signals shown in FIG. **3**, or by varying the time constant of the one-shots **74, 76** of FIG. **4**, so that this pulse width is provided by the signals of FIG. **5** at a lower oscillator frequency. In this way, the input coils are not left on longer than necessary. When either of the input coils is left on for a period of time longer than that necessary to produce the change in flux direction, power is being wasted through heating within the input coil without additional generation of power in the corresponding output coil.

A number of experiments have been conducted to determine the adequacy of an electromagnetic generator built as the generator **10** in FIG. **1** to produce power both to drive the switching and control logic, providing power to the input

US 6,362,718 B1

9

coils **26, 28**, and to drive an external load **44**. In the configuration used in this experiment, the input coils **26, 28** had 40 turns of 18-gauge copper wire, and the output coils **29, 30** had 450 turns of 18-gauge copper wire. The permanent magnet **12** had a height of 40 mm (1.575 in. between its north and south poles, in the direction of arrow **89**, a width of 25.4 mm (1.00 in.), in the direction of arrow **90**, and in the other direction, a depth of 38.1 mm (1.50 in.). The core **16** had a height, in the direction of arrow **89**, of 90 mm (3.542 in.), a width, in the direction of arrow **90**, of 135 mm (5.315 in.) and a depth of 70 mm (2.756 in.). The core **16** had a central hole with a height, in the direction of arrow **89**, of 40 mm (1.575 mm) to accommodate the magnet **12**, and a width, in the direction of arrow **90**, of 85 mm (3.346 in.). The core **16** was fabricated of two "C"-shaped halves, joined at lines **92**, to accommodate the winding of output coils **29, 30** and input coils **26, 28** over the core material.

The core material was a laminated iron-based magnetic alloy sold by Honeywell as METGLAS Magnetic Alloy 2605SA1. The magnet material was a combination of iron, neodymium, and boron.

The input coils **26, 28** were driven at an oscillator frequency of 87.5 KHz, which was determined to produce optimum efficiency using a switching control circuit configured as shown in FIG. **2**. This frequency has a period of 11.45 microseconds. The flip flop **54** is arranged, for example, to be set and reset on rising edges of the clock signal input from the oscillator, so that each pulse driving one of the FETS **60, 62** has a duration of 11.45 microseconds, and so that sequential pulses are also separated to each FET are also separated by 11.45 microseconds.

FIGS. **6A–6H** are graphical views of signals which simultaneously occurred within the apparatus of FIGS. **1** and **2** during operation with an applied input voltage of 75 volts. FIG. **6A** shows a first drive signal **100** driving FET **60**, which conducts to drive the right input coil **26**. FIG. **6B** is shows a second drive signal **102** driving FET **62**, which conducts to drive the left input coil **28**.

FIGS. **6C** and **6D** show voltage and current signals associated with current driving both the FETS **60, 62** from a battery source. FIG. **6C** shows the level **104** of voltage V. While the nominal voltage of the battery was 75 volts, a decaying transient signal **106** is superimposed on this voltage each time one of the FETS **60, 62** is switched on to conduct. The specific pattern of this transient signal depends on the internal resistance of the battery, as well as on a number of characteristics of the magnetic generator **10**. Similarly, FIG. **6D** shows the current **106** flowing into both FETS **60, 62** from the battery source. Since the signals **104, 106** show the effects of current flowing into both FETS **60, 62** the transient spikes are 11.45 microseconds apart.

FIGS. **6E–6H** show voltage and current levels measured at the output coils **29, 30**. FIG. **6E** shows a voltage output signal **108** of the right output coil **29**, while FIG. **6F** shows a voltage output signal **110** of the left output coil **30**. For example, the output current signal **116** of the right output coil **29** includes a first transient spike **112** caused when the a current pulse in the left input coil **28** is turned on to direct magnetic flux through the right magnetic path **18**, and a second transient spike **114** caused when the left input coil **28** is turned off with the right input coil **26** being turned on. FIG. **6G** shows a current output signal **116** of the right output coil **29**, while FIG. **6H** shows a current output signal **118** of the left output coil **30**.

FIG. **7** is a graphical view of output power measured using the electromagnetic generator **10** and eight levels of

10

input voltage, varying from 10v to 75v. The oscillator frequency was retained at 87.5 KHz. The measurement points are represented by indicia **120**, while the curve **122** is generated by polynomial regression analysis using a least squares fit.

FIG. **8** is a graphical view of a coefficient of performance, defined as the ratio of the output power to the input power, for each of the measurement points shown in FIG. **7**. At each measurement point, the output power was substantially higher than the input power. Real power measurements were computed at each data point using measured voltage and current levels, with the results being averaged over the period of the signal. These measurements agree with RMS power measured using a Textronic THS730 digital oscilloscope.

While the electromagnetic generator **10** was capable of operation at much higher voltages and currents without saturation, the input voltage was limited to 75 volts because of voltage limitations of the switching circuits being used. Those skilled in the relevant art will understand that components for switching circuits capable of handling higher voltages in this application are readily available. The experimentally-measured data was extrapolated to describe operation at an input voltage of 100 volts, with the input current being 140 ma, the input power being 14 watts, and with a resulting output power being 48 watts for each of the two output coils **29, 30**, at an average output current of 12 ma and an average output voltage of 4000 volts. This means that for each of the output coils **29, 30**, the coefficient of performance would be 3.44.

While an output voltage of 4000 volts may be needed for some applications, the output voltage can also be varied through a simple change in the configuration of the electromagnetic generator **10**. The output voltage is readily reduced by reducing the number of turns in the output windings. If this number of turns is decreased from 450 to 12, the output voltage is dropped to 106.7, with a resulting increase in output current to 0.5 amps for each output coil **29, 30**. In this way, the output current and voltage of the electromagnetic generator can be varied by varying the number of turns of the output coils **29, 30**, without making a substantial change in the output power, which is instead determined by the input current, which determines the amount of magnetic flux shuttled during the switching process.

The coefficients of performance, all of which were significantly greater than 1, plotted in FIG. **8** indicate that the output power levels measured in each of the output coils **29, 30** were substantially greater than the corresponding input power levels driving both of the input coils **26, 28**. Therefore, it is apparent that the electromagnetic generator **10** can be built in a self-actuating form, as discussed above in reference to FIG. **1**. In the example of FIG. **1**, except for a brief application of power from the external power source **38**, to start the process of power generation, the power required to drive the input coils **26, 28** is derived entirely from power developed within the right output coil **29**. If the power generated in a single output coil **29, 30** is more than sufficient to drive the input coils **26, 28**, an additional load **126** may be added to be driven with power generated in the output coil **29** used to generate power to drive the input coils **26, 28**. On the other hand, each of the output coils **29, 30** may be used to drive a portion of the input coil power requirements, for example with one of the output coils **26, 28** providing the voltage V for the FET **60** (shown in FIG. **2**), while the other output coil provides this voltage for the FET **62**.

Regarding thermodynamic considerations, it is noted that, when the electromagnetic generator **10** is operating, it is an

US 6,362,718 B1

11

open system not in thermodynamic equilibrium. The system receives static energy from the magnetic flux of the permanent magnet. Because the electromagnetic generator **10** is self-switched without an additional energy input, the thermodynamic operation of the system is an open dissipative system, receiving, collecting, and dissipating energy from its environment; in this case, from the magnetic flux stored within the permanent magnet. Continued operation of the electromagnetic generator **10** causes demagnetization of the permanent magnet. The use of a magnetic material including rare earth elements, such as a samarium cobalt material or a material including iron, neodymium, and boron is preferable within the present invention, since such a magnetic material has a relatively long life in this application.

Thus, an electromagnetic generator operating in accordance with the present invention should be considered not as a perpetual motion machine, but rather as a system in which flux radiated from a permanent magnet is converted into electricity, which is used both to power the apparatus and to power an external load. This is analogous to a system including a nuclear reactor, in which a number of fuel rods radiate energy which is used to keep the chain reaction going and to heat water for the generation of electricity to drive external loads.

FIG. 9 is a cross-sectional elevation of an electromagnetic generator **130** built in accordance with a second version of the first embodiment of the present invention. This electromagnetic generator **130** is generally similar in construction and operation to the electromagnetic generator **10** built in accordance with the first version of this embodiment, except that the magnetic core **132** of the electromagnetic generator **10** is built in two halves joined along lines **134**, allowing each of the output coils **135** to be wound on a plastic bobbin **136** before the bobbin **136** is placed over the legs **137** of the core **132**. FIG. 9 also shows an alternate placement of an input coil **138**. In the example of FIG. **1**, both input coils **26**, **28** were placed on the upper portion of the magnetic core **16**, with these coils **26**, **28** being configured to establish magnetic fields having north magnetic poles at the inner ends **31**, **32** of the coils **26**, **28**, with these north magnetic poles thus being closest to the end **14** of the permanent magnet **12** having its north magnetic pole. In the example of FIG. **9**, a first input coil **26** is as described above in reference to FIG. 1, but the second input coil **138** is placed adjacent the south pole **140** of the permanent magnet **12**. This input coil **138** is configured to establish a south magnetic pole at its inner end **142**, so that, when input coil **138** is turned on, flux from the permanent magnet **12** is directed away from the left magnetic path **20** into the right magnetic path **18**.

FIGS. **10** and **11** show an electromagnetic generator **150** built in accordance with a first version of a second embodiment of the present invention, with FIG. **10** being a top view thereof, and with FIG. **11** being a front elevation thereof. This electromagnetic generator **150** includes an output coil **152**, **153** at each corner, and a permanent magnet **154** extending along each side between output coils. The magnetic core **156** includes an upper plate **158**, a lower plate **160**, and a square post **162** extending within each output coil **152**, **153**. Both the upper plate **158** and the lower plate **160** include central apertures **164**.

Each of the permanent magnets **154** is oriented with a like pole, such as a north pole, against the upper plate **158**. Eight input coils **166**, **168** are placed in positions around the upper plate **158** between an output coil **152**, **153** and a permanent magnet **154**. Each input coil **166**, **168** is arranged to form a magnetic pole at its end nearest to the adjacent permanent magnet **154** of a like polarity to the magnetic poles of the

12

magnets **154** adjacent the upper plate **158**. Thus, the input coils **166** are switched on to divert magnetic flux of the permanent magnets **154** from the adjacent output coils **152**, with this flux being diverted into magnetic paths through the output coils **153**. Then, the input coils **168** are switched on to divert magnetic flux of the permanent magnets **154** from the adjacent output coils **153**, with this flux being diverted into magnetic paths through output coils **152**. Thus, the input coils form a first group of input coils **166** and a second group of input coils **168**, with these first and second groups of input coils being alternately energized in the manner described above in reference to FIG. **1** for the single input coils **26**, **28**. The output coils produce current in a first train of pulses occurring simultaneously within coils **152** and in a second train of pulses occurring simultaneously within coils **153**.

Thus, driving current through input coils **166** causes an increase in flux from the permanent magnets **154** within the posts **162** extending through output coils **153** and a decrease in flux from the permanent magnets **154** within the posts **162** extending through output coils **152**. On the other hand, driving current through input coils **168** causes a decrease in flux from the permanent magnets **154** within the posts **162** extending through output coils **153** and an increase in flux from the permanent magnets **154** within the posts **162** extending through output coils **152**.

While the example of FIGS. **10** and **11** shows all of the input coils **166,168** deployed along the upper plate **158**, it is understood that certain of these input coils **166**, **168** could alternately be deployed around the lower plate **160**, in the manner generally shown in FIG. **9**, with one input coil **166**, **168** being within each magnetic circuit between a permanent magnet **154** and an adjacent post **162** extending within an output coil **152**, **153**, and with each input coil **166**, **168** being arranged to produce a magnetic field having a magnetic pole like the closest pole of the adjacent permanent magnet **154**.

FIG. **12** is a top view of a second version **170** of the second embodiment of the present invention, which is similar to the first version thereof, which has been discussed in reference to FIGS. **10** and **11**, except that an upper plate **172** and a similar lower plate (not shown) are annular in shape, while the permanent magnets **174** and posts **176** extending through the output coils **178** are cylindrical. The input coils **180** are oriented and switched as described above in reference to FIGS. **9** and **10**.

While the example of FIG. **12** shows four permanent magnets, four output coils and eight input coils it is understood that the principles described above can be applied to electromagnetic generators having different numbers of elements. For example, such a device can be built to have two permanent magnets, two output coils, and four input coils, or to have six permanent magnets, six output coils, and twelve input coils.

In accordance with the present invention, material used for magnetic cores is preferably a nanocrystalline alloy, and alternately an amorphous alloy. The material is preferably in a laminated form. For example, the core material is a cobalt-niobium-boron alloy or an iron based magnetic alloy.

Also in accordance with the present invention, the permanent magnet material preferably includes a rare earth element. For example, the permanent magnet material is a samarium cobalt material or a combination of iron, neodymium, and boron.

While the invention has been described in its preferred versions and embodiments with some degree of particularity, it is understood that this description has been

US 6,362,718 B1

13

given only by way of example and that numerous changes in the details of construction, fabrication, and use, including the combination and arrangement of parts, may be made without departing from the spirit and scope of the invention.

What is claimed is:

1. An electromagnetic generator comprising:

a permanent magnet having magnetic poles at opposite ends;

a magnetic core including first and second magnetic paths between said opposite ends of said permanent magnet, wherein

said magnetic core comprises a closed loop,

said permanent magnet extends within said closed loop, and

said opposite ends of said permanent magnet are disposed adjacent opposite sides of said closed loop and against internal surfaces of said magnetic core comprising said closed loop;

a first input coil extending around a portion of said first magnetic path,

a second input coil extending around a portion of said second magnetic path,

a first output coil extending around a portion of said first magnetic path for providing a first electrical output;

a second output coil extending around a portion of said second magnetic path for providing a second electrical output; and

a switching circuit driving electrical current alternately through said first and second input coils, wherein

said electrical current driven through said first input coil causes said first input coil to produce a magnetic field opposing a concentration of magnetic flux from said permanent magnet within said first magnetic path, and

said electrical current driven through said second input coil causes said second input coil to produce a magnetic field opposing a concentration of magnetic flux from said permanent magnet within said second magnetic path.

2. An electromagnetic generator comprising:

a permanent magnet having magnetic poles at opposite ends;

a magnetic core including first and second magnetic paths between said opposite ends of said permanent magnet, wherein

said magnetic core comprises a closed loop,

said permanent magnet extends within said closed loop,

said opposite ends of said permanent magnet are disposed adjacent opposite sides of said closed loop, and

a first type of pole of said permanent magnet is disposed adjacent a first side of said closed loop;

a first input coil, disposed along said first side of said closed loop, extending around a portion of said first magnetic path,

a second input coil, disposed along said first side of said closed loop, extending around a portion of said second magnetic path,

a first output coil extending around a portion of said first magnetic path for providing a first electrical output;

a second output coil extending around a portion of said second magnetic path for providing a second electrical output; and

a switching circuit driving electrical current alternately through said first and second input coils, wherein

14

said electrical current driven through said first input coil causes said first input coil to produce a magnetic field opposing a concentration of magnetic flux from said permanent magnet within said first magnetic path, and additionally causes said first input coil to produce a magnetic field having said first type of pole at an end of said first input coil adjacent said permanent magnet, and

said electrical current driven through said second input coil causes said second input coil to produce a magnetic field opposing a concentration of magnetic flux from said permanent magnet within said second magnetic path, and additionally causes said second input coil to produce a magnetic field having said first type of pole at an end of said of said second input coil adjacent said permanent magnet.

3. An electromagnetic generator comprising:

a permanent magnet having magnetic poles at opposite ends;

a magnetic core including first and second magnetic paths between said opposite ends of said permanent magnet, wherein

said magnetic core comprises a closed loop,

said permanent magnet extends within said closed loop, and

said opposite ends of said permanent magnet are disposed adjacent opposite sides of said closed loop,

a first type of pole of said permanent magnet is disposed adjacent a first side of said closed loop, and

a second type of pole, opposite said first type of pole, of said permanent magnet is disposed adjacent a second side of said closed loop;

a first input coil extending around a portion of said first magnetic path, wherein said first input coil is disposed along said first side of said closed loop;

a second input coil extending around a portion of said second magnetic path wherein said second input coil is disposed along said second side of said closed loop;

a first output coil extending around a portion of said first magnetic path for providing a first electrical output;

a second output coil extending around a portion of said second magnetic path for providing a second electrical output; and

a switching circuit driving electrical current alternately through said first and second input coils, wherein

said electrical current driven through said first input coil causes said first input coil to produce a magnetic field opposing a concentration of magnetic flux from said permanent magnet within said first magnetic path, and additionally causes said first input coil to produce a magnetic field having said first type of pole at an end of said first input coil adjacent said permanent magnet, and

said electrical current driven through said second input coil causes said second input coil to produce a magnetic field opposing a concentration of magnetic flux from said permanent magnet within said second magnetic path, and additionally causes said second input coil to produce a magnetic field having said second type of pole at an end of said of said second input coil adjacent said permanent magnet.

4. An electromagnetic generator comprising:

a permanent magnet having magnetic poles at opposite ends;

a magnetic core including first and second magnetic paths between said opposite ends of said permanent magnet;

US 6,362,718 B1

15

a first input coil extending around a portion of said first magnetic path,

a second input coil extending around a portion of said second magnetic path,

a first output coil extending around a portion of said first magnetic path for providing a first electrical output;

a second output coil extending around a portion of said second magnetic path for providing a second electrical output; and

a switching circuit driving electrical current alternately through said first and second input coils, wherein said electrical current driven through said first input coil causes said first input coil to produce a magnetic field opposing a concentration of magnetic flux from said permanent magnet within said first magnetic path, and wherein said electrical current driven through said second input coil causes said second input coil to produce a magnetic field opposing a concentration of magnetic flux from said permanent magnet within said second magnetic path, wherein a portion of electrical power induced in said first output coil provides power to drive said switching circuit.

5. The electromagnetic generator of claim **4**, wherein said switching circuit is driven by an external power source during a starting process and by power induced in said first output coil during operation after said starting process.

6. The electromagnetic generator of claim **2**, wherein said magnetic core is composed of a nanocrystalline magnetic alloy.

7. The electromagnetic generator of claim **6**, wherein said nanocrystalline magnetic alloy is a cobalt-niobium-boron alloy.

8. The electromagnetic generator of claim **6**, wherein said nanocrystalline magnetic alloy is an iron-based alloy.

9. The electromagnetic generator of claim **2**, wherein said changes in flux density within said magnetic core occur without driving said magnetic core to magnetic saturation.

10. The electromagnetic generator of claim **2**, wherein

said switching circuit drives said electrical current through said first input coil in response to a first train of pulses,

said switching circuit drives said electrical current through said second input coil in response to a second train of pulses, alternating with pulses within said first train of pulses, and

said pulses in said first and second trains of pulses are approximately 11.5 milliseconds in duration.

11. The electromagnetic generator of claim **2**, wherein said permanent magnet is composed of a material including a rare earth element.

12. The electromagnetic generator of claim **11**, wherein said permanent magnet is composed essentially of samarium cobalt.

13. The electromagnetic generator of claim **11**, wherein said permanent magnet is composed essentially of iron, neodymium, and boron.

14. An electromagnetic generator comprising:

a magnetic core including a pair of spaced-apart plates, wherein each of said spaced-apart plates includes a central aperture, and first and second pluralities of posts extending between said spaced-apart plates;

a plurality of permanent magnets extending individually between said pair of spaced-apart plates and between adjacent posts within said plurality of posts, wherein each permanent magnet within said plurality of permanent magnets has magnetic poles at opposite ends,

16

wherein all magnets within said plurality of magnets are oriented to produce magnetic fields having a common direction;

first and second pluralities of input coils, wherein each input coil within said first and second pluralities of input coils extends around a portion of a plate within said spaced-apart plates between a post in said plurality of posts and a permanent magnet in said plurality of permanent magnets;

an output coil extending around each post in said first and second pluralities of posts for providing an electrical output;

a switching circuit driving electrical current alternatively through said first and second pluralities of input coils, wherein said electrical current driven through each input coil in said first plurality of input coils causes an increase in magnetic flux within each post within said first plurality of posts from permanent magnets on each side of said post and a decrease in magnetic flux within each post within said second plurality of posts from permanent magnets on each side of said post, and wherein said electrical current driven through input coil in said second plurality of input coils causes a decrease in magnetic flux within each post within said first plurality of posts from permanent magnets on each side of said post and an increase in magnetic flux within each post within said second plurality of posts from permanent magnets on each side of said post.

15. The electromagnetic generator of claim **14**, wherein

each input coil extends around a portion of a magnetic path through said magnetic core between said opposite ends a permanent magnet adjacent said input coil,

said magnetic path extends through a post within said magnetic core adjacent said input coil, and

driving electrical current through said input coil causes said input coil to produce a magnetic field opposing a concentration of magnetic flux within said magnetic path.

16. The electromagnetic generator of claim **14**, wherein said switching circuit is driven by an external power source during a starting process and by power induced in said output coils during operation after said starting process.

17. The electromagnetic generator of claim **14**, wherein said magnetic core is composed of a nanocrystalline magnetic alloy.

18. The electromagnetic generator of claim **2**, wherein a portion of electrical power induced in said first output coil provides power to drive said switching circuit.

19. The electromagnetic generator of claim **18**, wherein said switching circuit is driven by an external power source during a starting process and by power induced in said first output coil during operation after said starting process.

20. The electromagnetic generator of claim **3**, wherein a portion of electrical power induced in said first output coil provides power to drive said switching circuit.

21. The electromagnetic generator of claim **20**, wherein said switching circuit is driven by an external power source during a starting process and by power induced in said first output coil during operation after said starting process.

22. The electromagnetic generator of claim **3**, wherein said magnetic core is composed of a nanocrystalline magnetic alloy.

23. The electromagnetic generator of claim **22**, wherein said nanocrystalline magnetic alloy is a cobalt-niobium-boron alloy.

US 6,362,718 B1

17

24. The electromagnetic generator of claim **22**, wherein said nanocrystalline magnetic alloy is an iron-based alloy.

25. The electromagnetic generator of claim **3**, wherein said changes in flux density within said magnetic core occur without driving said magnetic core to magnetic saturation.

26. The electromagnetic generator of claim **3**, wherein

said switching circuit drives said electrical current through said first input coil in response to a first train of pulses,

said switching circuit drives said electrical current through said second input coil in response to a second train of pulses, alternating with pulses within said first train of pulses, and

18

said pulses in said first and second trains of pulses are approximately 11.5 milliseconds in duration.

27. The electromagnetic generator of claim **3**, wherein said permanent magnet is composed of a material including a rare earth element.

28. The electromagnetic generator of claim **27**, wherein said permanent magnet is composed essentially of samarium cobalt.

29. The electromagnetic generator of claim **27**, wherein said permanent magnet is composed essentially of iron, neodymium, and boron.

* * * * *

Chapter 4
Anti-Gravity Propulsion Dynamics
A Brief Overview
by
Paul Potter

FROM:

ANTI-GRAVITY PROPULSION DYNAMICS
UFOs and Gravitational Manipulation, New Edition
By Paul E. Potter

Paul Potter's book introduces a brand new field of scientific research based upon analysis of artifacts retrieved from crashed and damaged UFOs that have come down in Russia and America. For the first time, it reveals the scientific principles behind UFO propulsion dynamics, and shows that these principles are known and recognized by today's physicists. Potter's analyses of these UFO mechanisms are substantiated with references to a broad array of over 300 research papers published in scientific journals! Potter correlates many of the phenomena observed firsthand by close encounter witnesses and abductees and pinpoints the common themes reported, categorizing them according to known physical principles. He produces a comprehensive orchestration of energy dynamics used inside and around UFOs. His precise and lavish illustrations allow the reader to enter directly into the realm of the advanced technological engineer and to understand, quite straightforwardly, the aliens' methods of energy manipulation: their methods of electrical power generation; how they purposely designed their craft to employ the kinds of energy dynamics that are exclusive to space (discoverable in our astrophysics) in order that their craft may generate both attractive and repulsive gravitational forces; their control over the mass-density matrix surrounding their craft enabling them to alter their physical dimensions and even manufacture their own frame of reference in respect to time.
Available from Adventures Unlimited Press and Amazon.com

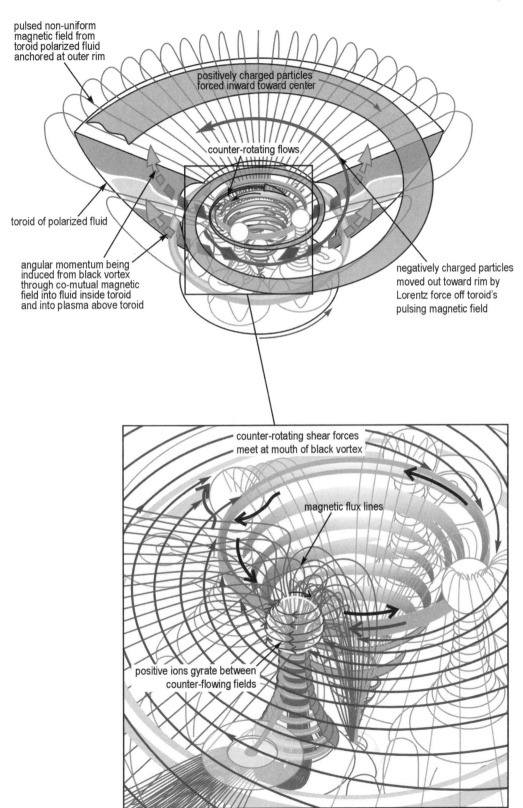

pulsed non-uniform magnetic field from toroid polarized fluid anchored at outer rim

positively charged particles forced inward toward center

counter-rotating flows

toroid of polarized fluid

angular momentum being induced from black vortex through co-mutual magnetic field into fluid inside toroid and into plasma above toroid

negatively charged particles moved out toward rim by Lorentz force off toroid's pulsing magnetic field

counter-rotating shear forces meet at mouth of black vortex

magnetic flux lines

positive ions gyrate between counter-flowing fields

61

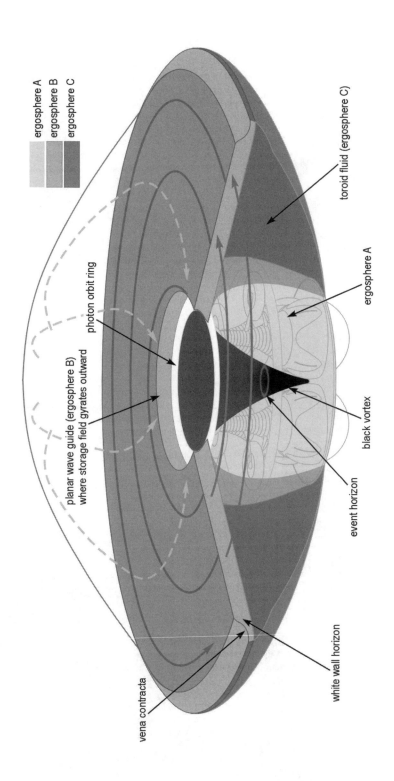

ergosphere A
ergosphere B
ergosphere C

toroid fluid (ergosphere C)

ergosphere A

black vortex

event horizon

photon orbit ring

planar wave guide (ergosphere B)
where storage field gyrates outward

white wall horizon

vena contracta

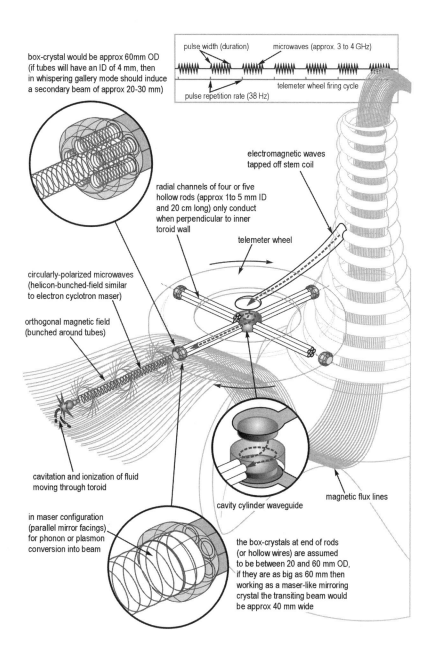

box-crystal would be approx 60mm OD
(if tubes will have an ID of 4 mm, then
in whispering gallery mode should induce
a secondary beam of approx 20-30 mm)

pulse width (duration)   microwaves (approx. 3 to 4 GHz)

telemeter wheel firing cycle

pulse repetition rate (38 Hz)

electromagnetic waves
tapped off stem coil

radial channels of four or five
hollow rods (approx 1to 5 mm ID
and 20 cm long) only conduct
when perpendicular to inner
toroid wall

telemeter wheel

circularly-polarized microwaves
(helicon-bunched-field similar
to electron cyclotron maser)

orthogonal magnetic field
(bunched around tubes)

cavitation and ionization of fluid
moving through toroid

magnetic flux lines

in maser configuration
(parallel mirror facings)
for phonon or plasmon
conversion into beam

cavity cylinder waveguide

the box-crystals at end of rods
(or hollow wires) are assumed
to be between 20 and 60 mm OD,
if they are as big as 60 mm then
working as a maser-like mirroring
crystal the transiting beam would
be approx 40 mm wide

iron-ferroelectric balls

wire-quartz network

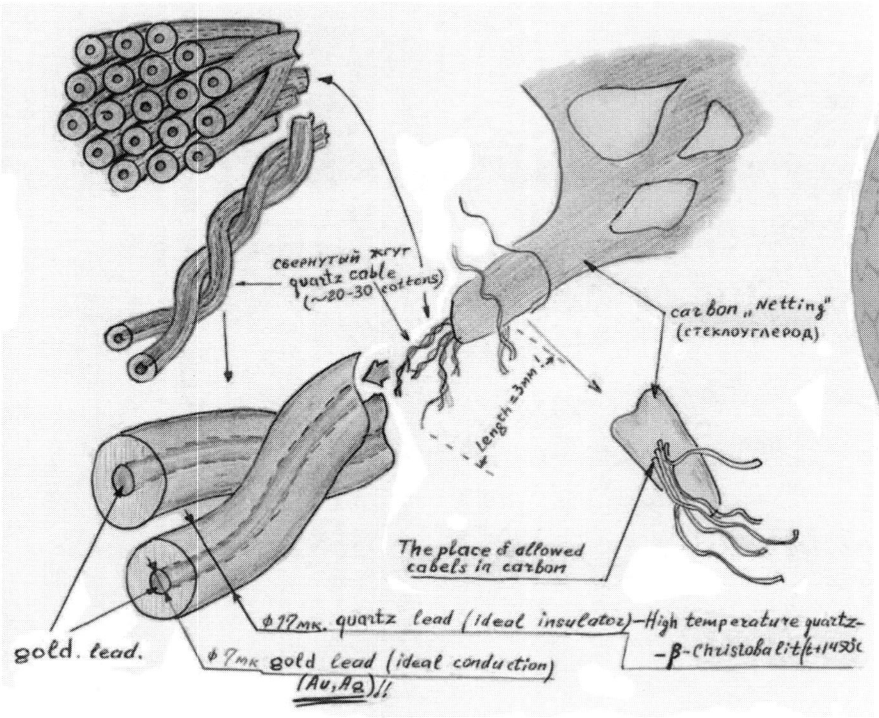

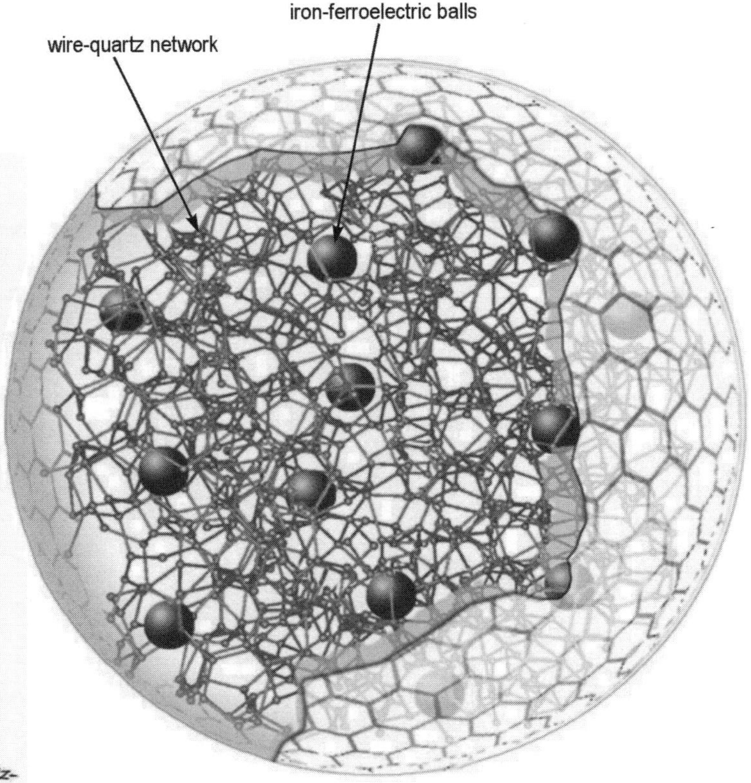

as the wire network might appear inside a dielectric sphere together with iron-rare-earth balls embedded into the mesh (these iron-rare-earth balls are enlarged for clarity)

Dalnegorsk UFO wire-quartz mesh drawings
(from 80x optical microscope photographs)
copyright © 2008 Valeri Dvuzhilny

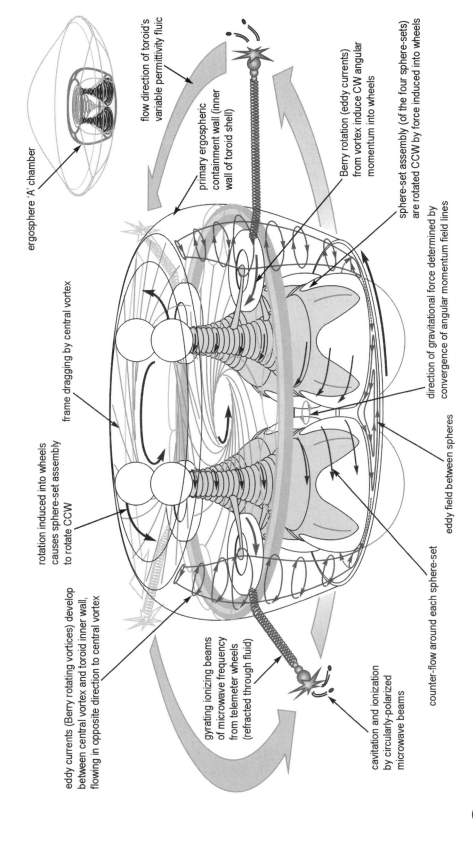

es

ergosphere 'A' chamber

flow direction of toroid's variable permittivity fluic

primary ergospheric containment wall (inner wall of toroid shell)

Berry rotation (eddy currents) from vortex induce CW angular momentum into wheels

sphere-set assembly (of the four sphere-sets) are rotated CCW by force induced into wheels

direction of gravitational force determined by convergence of angular momentum field lines

eddy field between spheres

counter-flow around each sphere-set

frame dragging by central vortex

rotation induced into wheels causes sphere-set assembly to rotate CCW

eddy currents (Berry rotating vortices) develop between central vortex and toroid inner wall, flowing in opposite direction to central vortex

gyrating ionizing beams of microwave frequency from telemeter wheels (refracted through fluid)

cavitation and ionization by circularly-polarized microwave beams

65

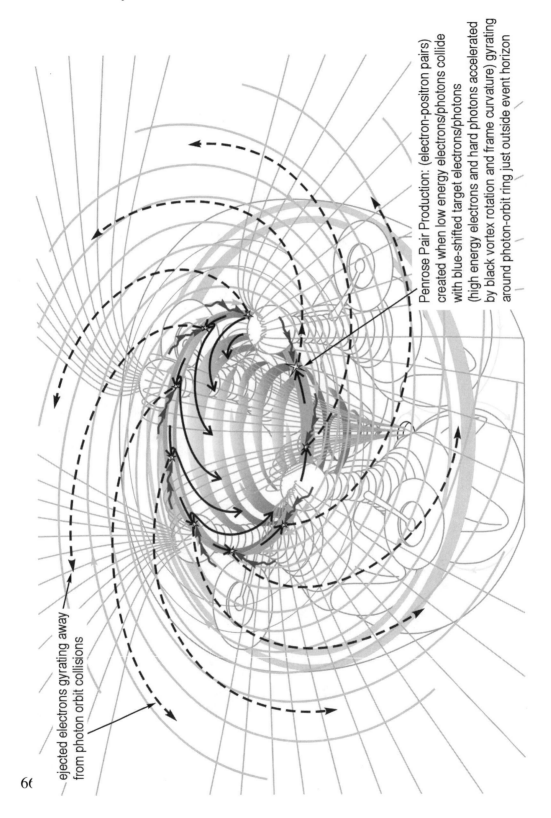

Penrose Pair Production: (electron-positron pairs) created when low energy electrons/photons collide with blue-shifted target electrons/photons (high energy electrons and hard photons accelerated by black vortex rotation and frame curvature) gyrating around photon-orbit ring just outside event horizon

ejected electrons gyrating away from photon orbit collisions

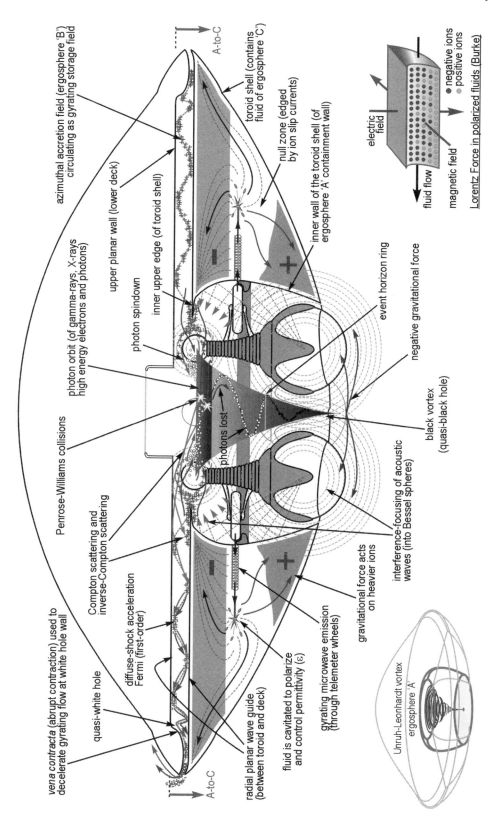

67

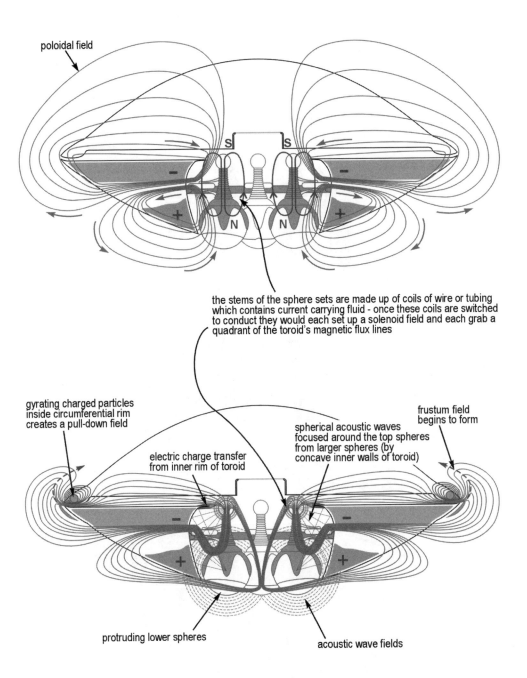

poloidal field

the stems of the sphere sets are made up of coils of wire or tubing which contains current carrying fluid - once these coils are switched to conduct they would each set up a solenoid field and each grab a quadrant of the toroid's magnetic flux lines

gyrating charged particles inside circumferential rim creates a pull-down field

frustum field begins to form

spherical acoustic waves focused around the top spheres from larger spheres (by concave inner walls of toroid)

electric charge transfer from inner rim of toroid

protruding lower spheres

acoustic wave fields

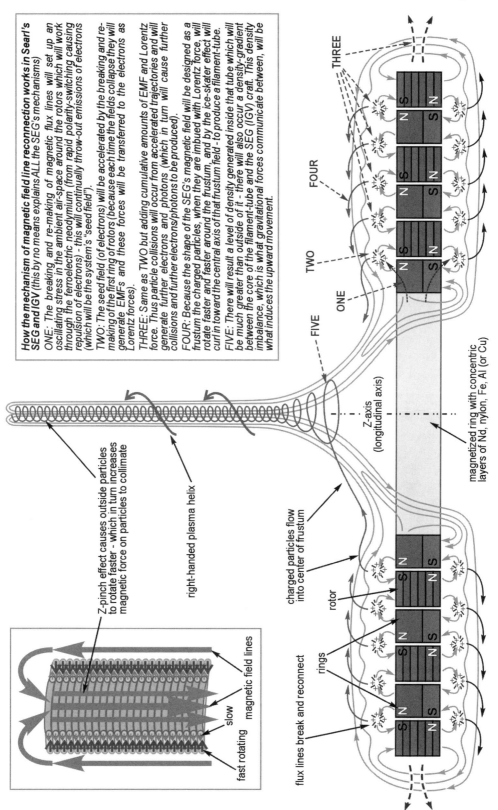

How the mechanism of magnetic field line reconnection works in Searl's SEG and IGV (this by no means explains ALL the SEG's mechanisms)

ONE: The breaking and re-making of magnetic flux lines will set up an oscillating stress in the ambient air-space around the rotors which will work through the ferroelectric neodymium (from rapid polarity-switching causing repulsion of electrons) - this will continually throw-out emissions of electrons (which will be the system's "seed field").

TWO: The seed field (of electrons) will be accelerated by the breaking and re-making of the first ring of rotors (because each time the fields collapse they will generate EMFs and these forces will be transferred to the electrons as Lorentz forces).

THREE: Same as TWO but adding cumulative amounts of EMF and Lorentz force. Thus particle collisions will occur from accelerated trajectories and will generate further electrons and photons (which in turn will cause further collisions and further electrons/photons to be produced).

FOUR: Because the shape of the SEG's magnetic field will be designed as a frustum the charged particles, when they are imbued with Lorentz force, will rotate faster and faster around the frustum, and by the ice-skater effect will curl in toward the central axis of that frustum field - to produce a filament-tube.

FIVE: There will result a level of density generated inside that tube which will be much greater than outside of it - there will also occur a density-gradient between the core of the filament-tube and the SEG (IGV) craft. This density imbalance, which is what gravitational forces communicate between, will be what induces the upward movement.

THREE

FOUR

TWO

ONE

FIVE

Z-axis
(longitudinal axis)

magnetized ring with concentric
layers of Nd, nylon, Fe, Al (or Cu)

Z-pinch effect causes outside particles
to rotate faster - which in turn increases
magnetic force on particles to collimate

right-handed plasma helix

charged particles flow
into center of frustum

rotor

rings

flux lines break and reconnect

slow   magnetic field lines

fast rotating

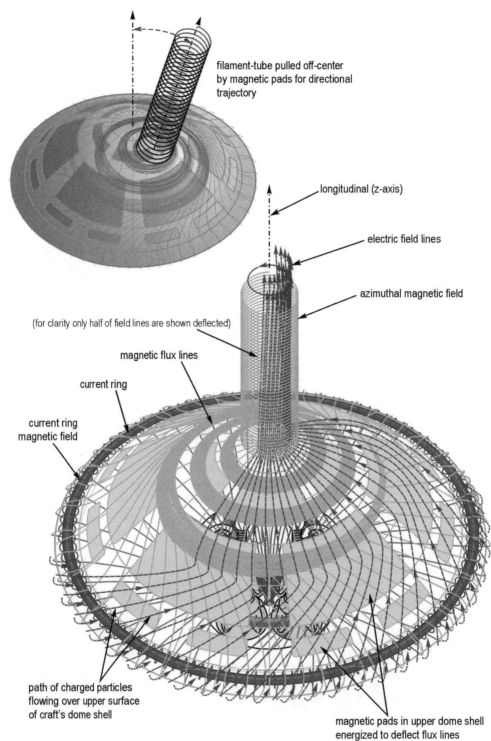

filament-tube pulled off-center
by magnetic pads for directional
trajectory

longitudinal (z-axis)

electric field lines

azimuthal magnetic field

(for clarity only half of field lines are shown deflected)

magnetic flux lines

current ring

current ring
magnetic field

path of charged particles
flowing over upper surface
of craft's dome shell

magnetic pads in upper dome shell
energized to deflect flux lines

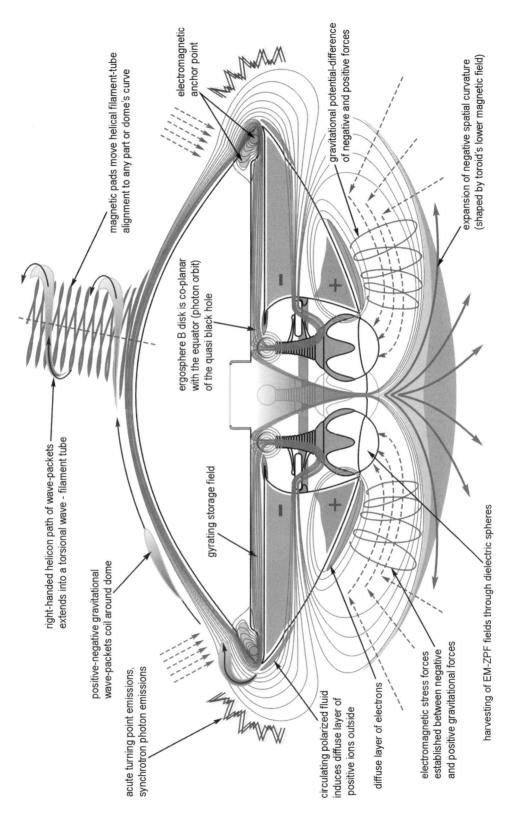

electromagnetic anchor point

magnetic pads move helical filament-tube alignment to any part or dome's curve

gravitational potential-difference of negative and positive forces

expansion of negative spatial curvature (shaped by toroid's lower magnetic field)

right-handed helicon path of wave-packets extends into a torsional wave - filament tube

positive-negative gravitational wave-packets coil around dome

acute turning point emissions, synchrotron photon emissions

ergosphere B disk is co-planar with the equator (photon orbit) of the quasi black hole

gyrating storage field

circulating polarized fluid induces diffuse layer of positive ions outside

diffuse layer of electrons

electromagnetic stress forces established between negative and positive gravitational forces

harvesting of EM-ZPF fields through dielectric spheres

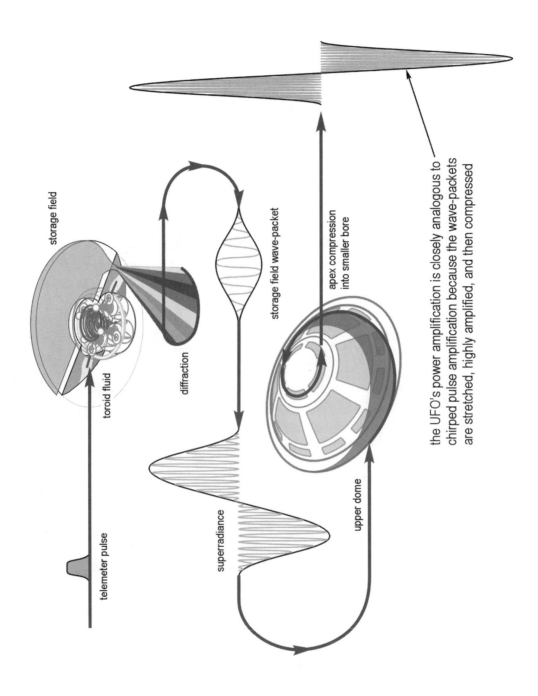

storage field

toroid fluid

telemeter pulse

diffraction

superradiance

storage field wave-packet

apex compression into smaller bore

upper dome

the UFO's power amplification is closely analogous to chirped pulse amplification because the wave-packets are stretched, highly amplified, and then compressed

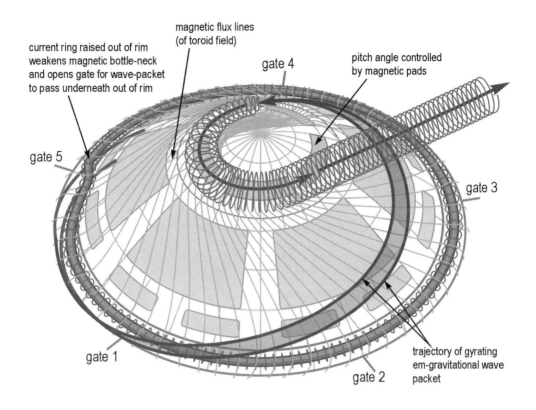

current ring raised out of rim weakens magnetic bottle-neck and opens gate for wave-packet to pass underneath out of rim

magnetic flux lines (of toroid field)

gate 4

pitch angle controlled by magnetic pads

gate 5

gate 3

gate 1

gate 2

trajectory of gyrating em-gravitational wave packet

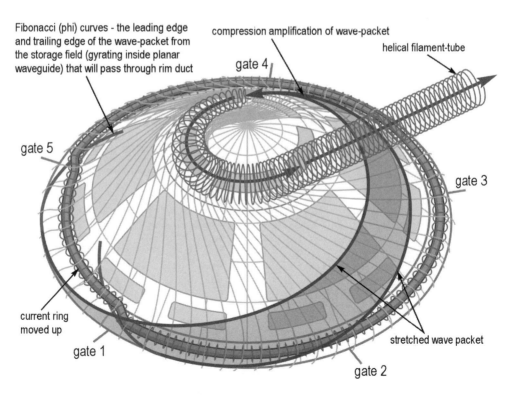

Fibonacci (phi) curves - the leading edge and trailing edge of the wave-packet from the storage field (gyrating inside planar waveguide) that will pass through rim duct

compression amplification of wave-packet

helical filament-tube

gate 4

gate 5

gate 3

current ring moved up

gate 1

gate 2

stretched wave packet

73

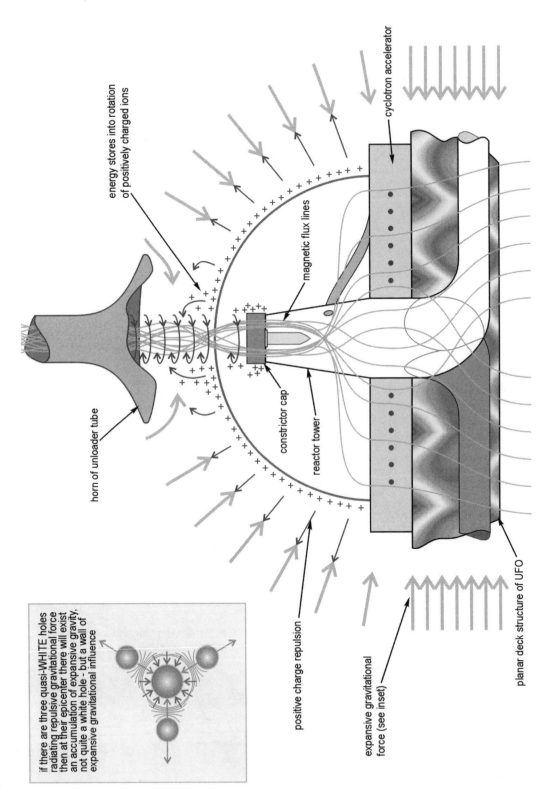

cyclotron accelerator

energy stores into rotation of positively charged ions

magnetic flux lines

horn of unloader tube

constrictor cap

reactor tower

planar deck structure of UFO

positive charge repulsion

expansive gravitational force (see inset)

if there are three quasi-WHITE holes radiating repulsive gravitational force then at their epicenter there will exist an accumulation of expansive gravity, not quite a white hole - but a wall of expansive gravitational influence

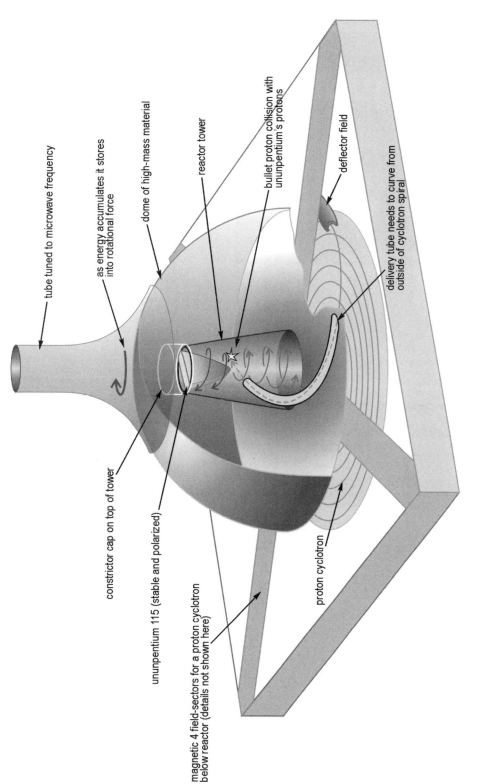

tube tuned to microwave frequency

as energy accumulates it stores into rotational force

dome of high-mass material

reactor tower

bullet proton collision with ununpentium's protons

deflector field

delivery tube needs to curve from outside of cyclotron spiral

proton cyclotron

constrictor cap on top of tower

ununpentium 115 (stable and polarized)

magnetic 4 field-sectors for a proton cyclotron below reactor (details not shown here)

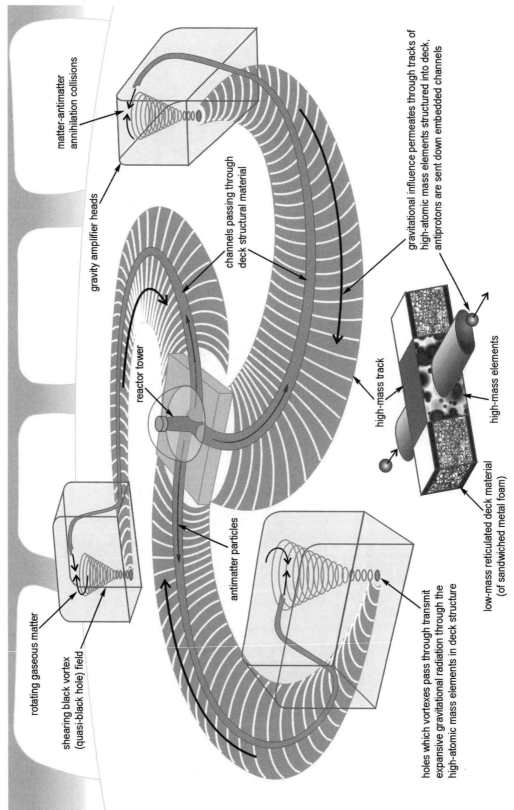

matter-antimatter annihilation collisions

gravity amplifier heads

channels passing through deck structural material

gravitational influence permeates through tracks of high-atomic mass elements structured into deck, antiprotons are sent down embedded channels

reactor tower

high-mass track

high-mass elements

antimatter particles

rotating gaseous matter

shearing black vortex (quasi-black hole) field

low-mass reticulated deck material (of sandwiched metal foam)

holes which vortexes pass through transmit expansive gravitational radiation through the high-atomic mass elements in deck structure

Chapter 4
More
Anti-Gravity Patents

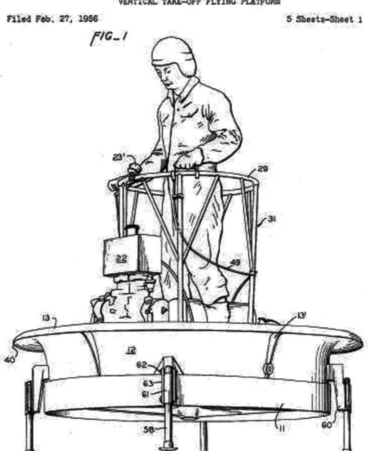

Sept. 20, 1960   A. C. ROBERTSON ET AL   2,953,321

VERTICAL TAKE-OFF FLYING PLATFORM

Filed Feb. 27, 1956   5 Sheets-Sheet 1

FIG_1

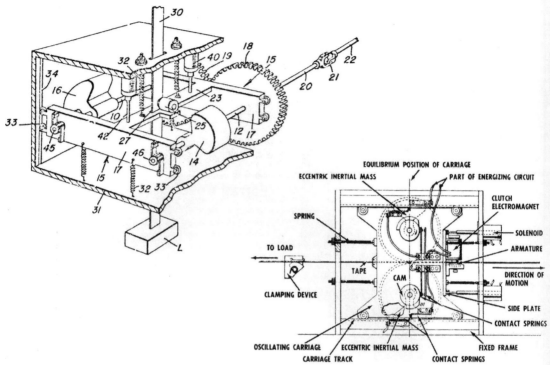

Dean drive.

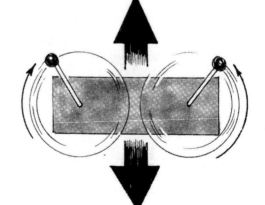

The heart of Dean's mechanism.

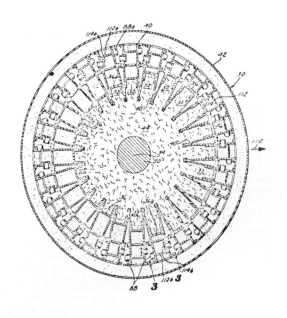

Matyas drive.

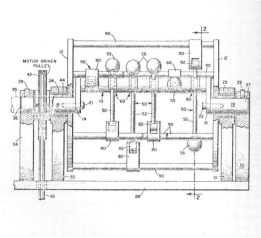

Novak drive.

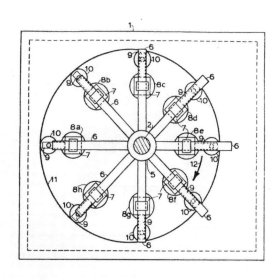

Cuff drive.

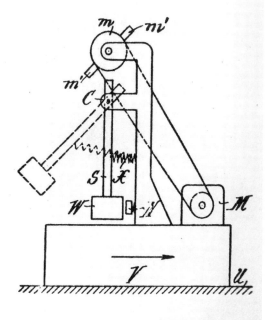

Goldschmidt drive.

79

Michel Wibault´s
Aircraft with Enclosed Rotor
U.S.-patent 2,807,428, Sept. 24, 1957

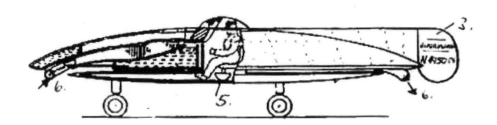

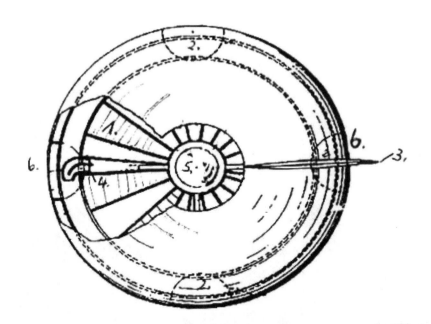

1. Centrifugal Air Compressor/Impeller	4. Ducted Burner, similar to tip-jets at heli-blades
2. Fluid Ballast Tanks, interconnected	5. Cockpit in the middle of the Hub
3. Fin for stability reasons	6. Air intake in front and nozzles at the rear

The large rotating impeller works as a gyroscope too, to stabilizy the whole craft in flight.

Rotating Disc Aircraft
U.S.-Patent 4,214,720, July 29, 1980

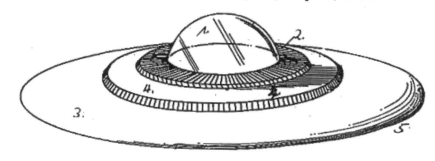

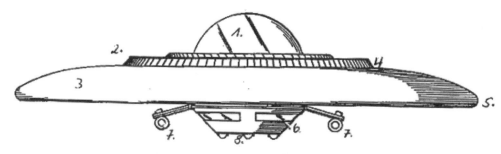

1. Cockpit	5. Trailing Edge
2. Cover Surface	6. Lower Cockpit Housing
3. Disc-Wing	7. Horizontal Flying Means, Jet Engines
4. Turbine Blades	8. Landing-Gear

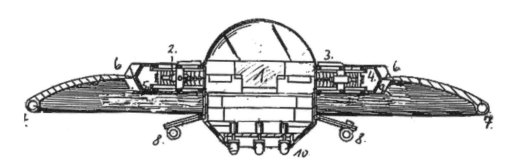

1. Cockpit, Seat and Consoles
2. Thrust Means (8 sets of two), Jet-Engines, radially mounted adjacent and around the Ceockpit
3. Cover Surface
4. Turbine Blades
5. Second Set of Turbine Baldes
6. Leading edge
7. Trailing Edge
8. Engines producing thrust for horizontal flight, two on each side
9. Rudder (4), Thrust deverting stabilizing menas, holding cockpit section in place
10. Three wheels, retractable

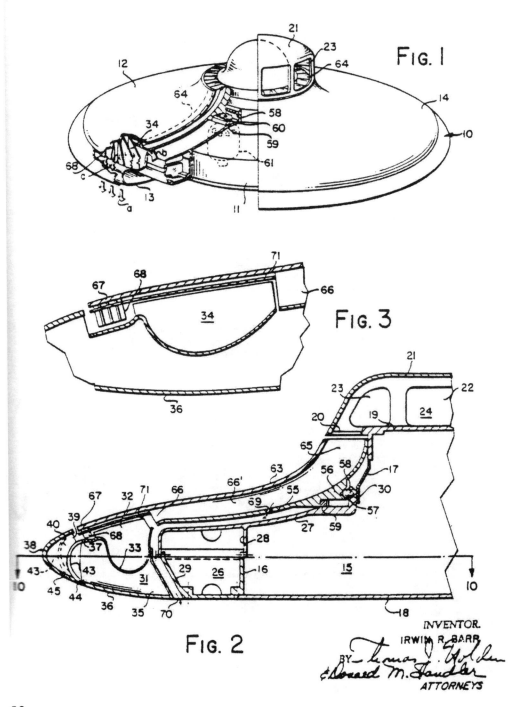

FIG. 1

FIG. 3

FIG. 2

INVENTOR.
IRWIN R. BARR
BY
ATTORNEYS

82

Dec. 11, 1962

Filed Nov. 19, 1958

I. R. BARR

FLYING MACHINE

3,067,967

3 Sheets—Sheet 2

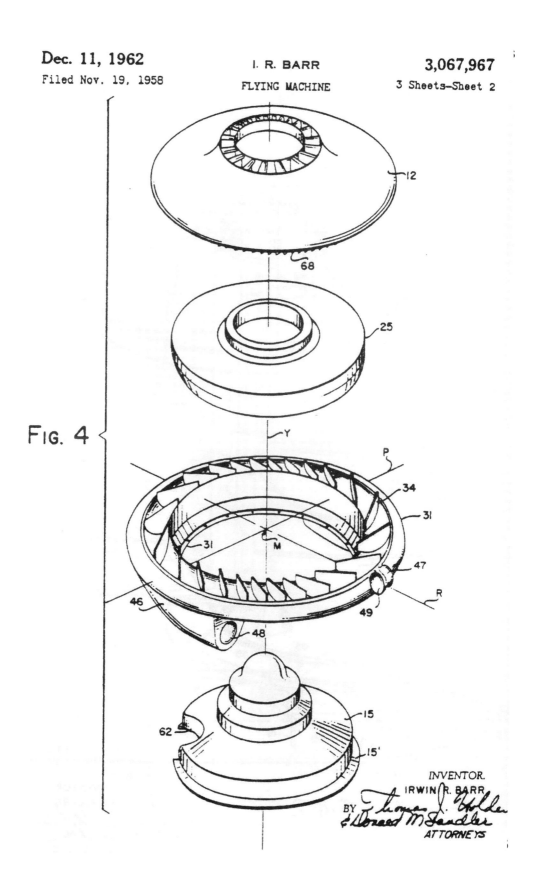

FIG. 4

Gyro Stabilized Vertical Rising Vehicle
(Discoid)

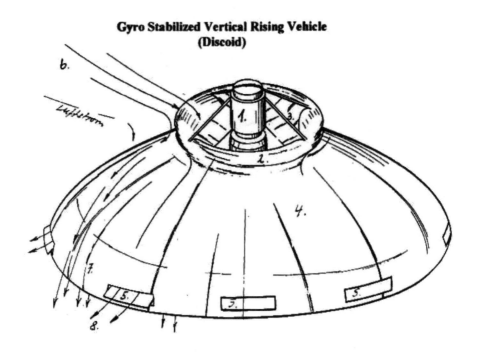

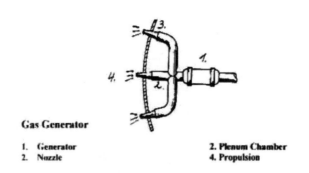

Gas Generator

1. Generator
2. Nozzle

2. Plenum Chamber
4. Propulsion

The Inventors T.P. Mulgrave and the German mathematician Dr. Friedrich O. Ringleb, who was a close assistant to Dr. Lippisch, working for LFW, Vienna, designed this impeller-saucer.

The vehicle could possibly be based on a former German design (see the „Ducted-Fan Saucer") which Dr. Ringleb brought to the USA after the war.

Gyro Stabilized Vertical Rising Vehicle
(Discoid)

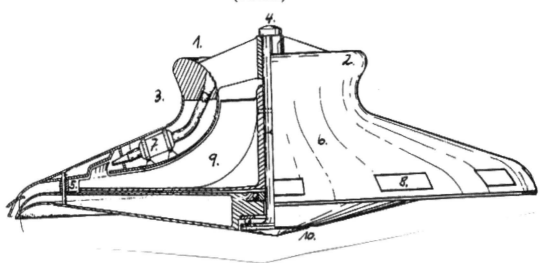

1. Central circular Opening	6. Bell-shaped Wing
2. „Crown" or „Lip"	7. Gas Turbine with Nozzle
3. S-shaped Contour	8. Variable Louver
4. Shaft	9. Impeller Blade
5. Vane	10. Bell-shaped Belly

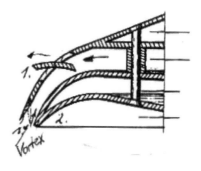

Peripherical Lip

1. Louver

2. Cusp Configuration

1. rotating central Shaft	5. Outlet
2. „Crown" or „Lip"	6. Air coming inwardly
3. Ribs	7. Outwardly Airflow
4. Bell-shaped outer section	8. Air/Exhaust from Impeller

March 11, 1969   E. GUERRERO   3,432,120

AIRCRAFT

Filed May 20, 1966   Sheet / of 5

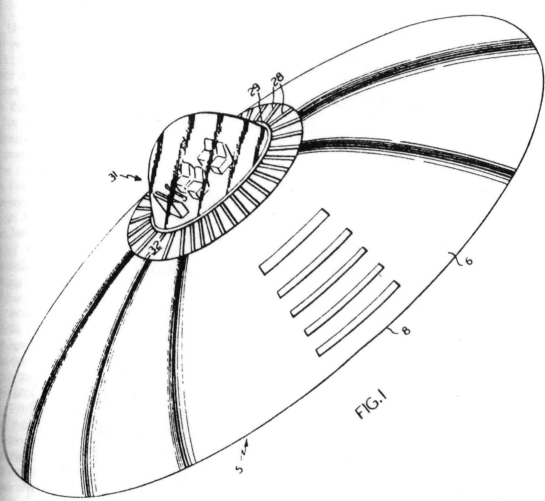

FIG.1

INVENTOR
EFRAIN GUERRERO

BY *Philpitt, Steininger & Preddy*
ATTORNEYS

March 11, 1969

Filed May 20, 1966

E. GUERRERO

AIRCRAFT

3,432,120

Sheet 2 of 3

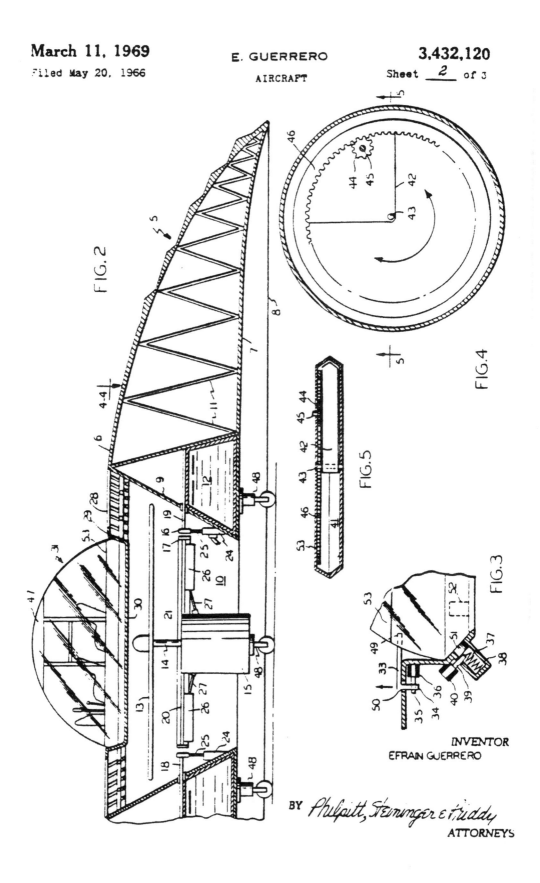

FIG.2

FIG.4

FIG.5

FIG.3

INVENTOR

EFRAIN GUERRERO

BY *Philpitt, Steininger & Huddy*

ATTORNEYS

[54] **PULSED CAPACITOR DISCHARGE ELECTRIC ENGINE**

[75] Inventor: Edwin V. Gray, Northridge, Calif.

[73] Assignee: Evgray Enterprises, Inc., Van Nuys, Calif.

[22] Filed: Nov. 2, 1973

[21] Appl. No.: 412,415

[52] U.S. Cl. 318/139; 318/254; 318/439; 310/46

[51] Int. Cl. H02p 5/00

[58] Field of Search 310/46, 5, 6; 318/194, 318/439, 254, 139; 320/1; 307/110

[56]   **References Cited**
             UNITED STATES PATENTS

2,085,708   6/1937   Spencer 318/194
2,800,619   7/1957   Brunt 318/194
3,579,074   5/1971   Roberts............................ 320/1
3,619,638   11/1971 Phinney 307/110

            OTHER PUBLICATIONS

Frungel, *High Speed Pulse Technology*, Academic Press Inc., 1965, pp. 140-148.

Primary Examiner—Robert K. Schaefer
Assistant Examiner—John J. Feldhaus
Attorney, Agent, or Firm—Gerald L. Price

[57]   **ABSTRACT**

There is disclosed herein an electric machine or engine in which a rotor cage having an array of electromagnets is rotatable in an array of electromagnets, or fixed electromagnets are juxtaposed against movable ones. The coils of the electromagnets are connected in the discharge path of capacitors charged to relatively high voltage and discharged through the electromagnetic coils when selected rotor and stator elements are in alignment, or when the fixed electromagnets and movable electromagnets are juxtaposed. The discharge occurs across spark gaps disclosed in alignment with respect to the desired juxtaposition of the selected movable and stationary electromagnets. The capacitor discharges occur simultaneously through juxtaposed stationary movable electromagnets wound so that their respective cores are in magnetic repulsion polarity, thus resulting in the forced motion of movable electromagnetic elements away from the juxtaposed stationary electromagnetic elements at the discharge, thereby achieving motion. In an engine, the discharges occur successively across selected ones of the gaps to maintain continuous rotation. Capacitors are recharged between successive alignment positions of particular rotor and stator electromagnets of the engine.

18 Claims, 19 Drawing Figures

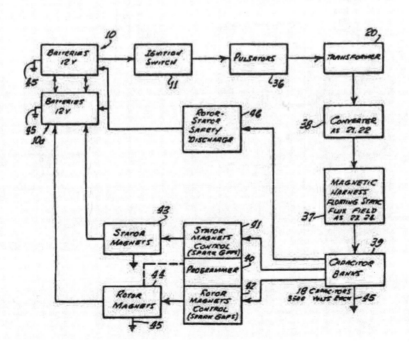

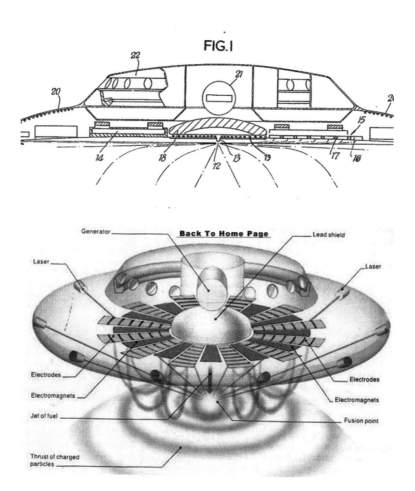

The Britrail design for a superconducting-magnetic saucer.

Sept. 4, 1951
A. S. TREMULIS
AUTOMOBILE HOOD ORNAMENT
OR SIMILAR ARTICLE
Filed Jan. 27, 1951
Des. 164,461

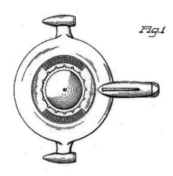

Fig.1

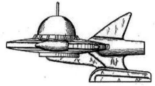

Fig.2

Fig.3

Inventor:-
Alexander S. Tremulis
by Hill, Sherman, Meroni, Gross & Simpson
Attys

Feb. 8, 1949.
C. A. NEUMANN ET AL
2,461,435
FLUID PROPELLED AND CONTROLLED AIRCRAFT
Filed April 19, 1945
5 Sheets—Sheet 1

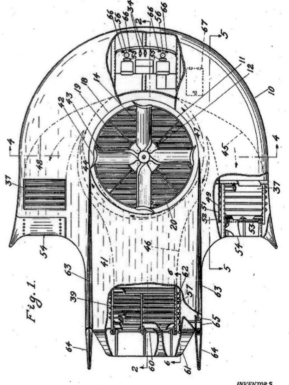

Fig. 1.

INVENTORS
C.A. Neumann & H.H. Baca.
BY
A.D.Adams
ATTORNEY.

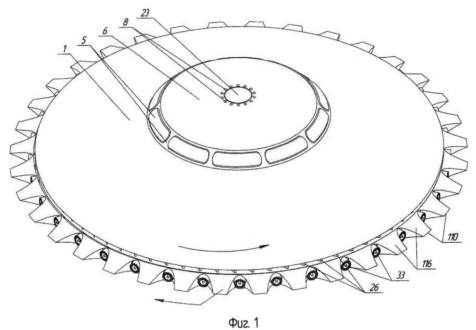

Фиг. 1

The latest Russian design for a flying saucer according to RF News.

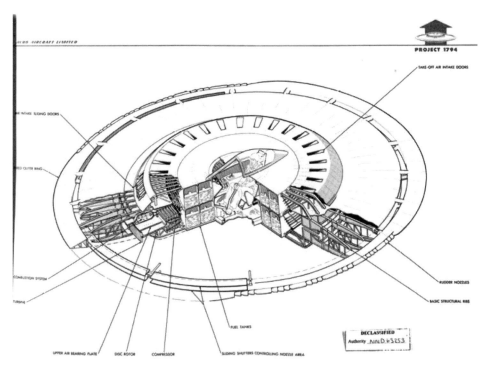

Project 794, now declassified.

The Anti-Gravity Files

Chapter 5

The Machines in Flight

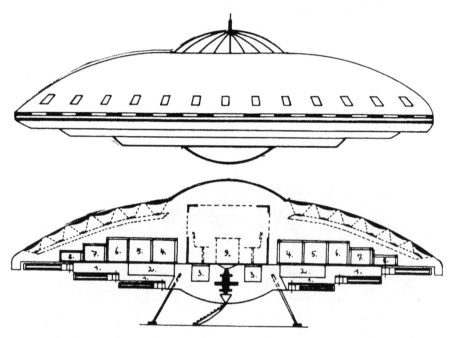

Diagrams of the Haunebu craft allegedly built by the Germans during WWII.

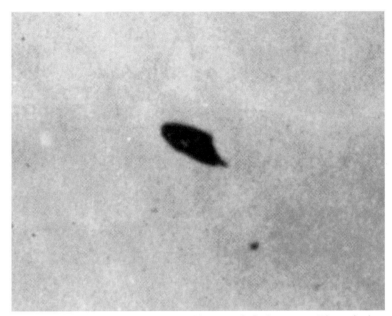

A daylight photo of an unidentified craft flying over Phoenix in
July of 1947.

A daylight photo of an unidentified craft flying
over Utah on June 10, 1964

A daylight photo of an unidentified craft flying
over Ward, Colorado in April of 1929.

The Anti-Gravity Files

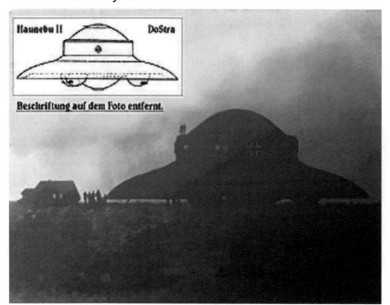

A photo and diagrams of the Haunebu craft allegedly built by the Germans during WWII.

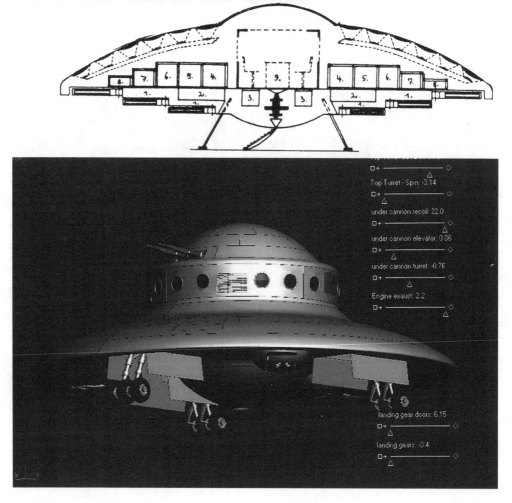

A very clear photo of a flying disc taken near the beach town of Floridad, Uruguay on July 11, 1977.

Right: A rare color photo of an alleged Vril craft built by the Germans during WWII.

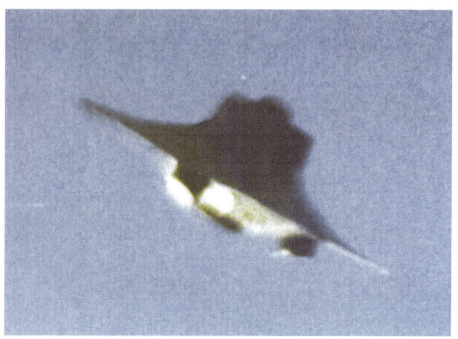

An astonishing series of photos taken somewhere in the North Atlantic by the US Navy submarine *Trepang* in March of 1971. They show a large tubular UFO emerging from the ocean and flying away.

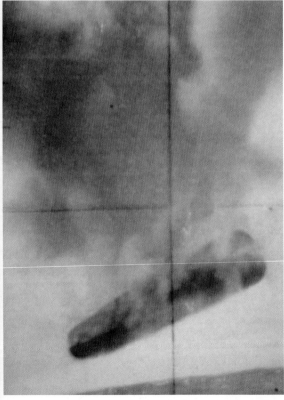

More of the astonishing series of photos taken somewhere in the North Atlantic by the US Navy submarine *Trepang* in March of 1971. *Below*: Two more photos taken on the same day by the *USS Trepang* of a different craft exiting the ocean, this one smaller and more like a disc or triangle.

Indiana-01-31-08

Left: A photo of a tubular craft taken in Indiana on the night of January 31, 2008. *Below Right*: A fascinating series of photos taken over LaGrange, Oregon on August 15, 2013.

Below Left: An astonishing series of photos of an arrowhead-shaped UFO hovering in the summer skies of Dayton, Ohio in May of 2016. Dayton is the location of the important Wright-Patterson Air Force Base.

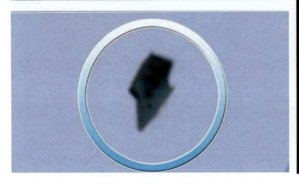

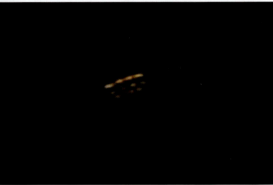

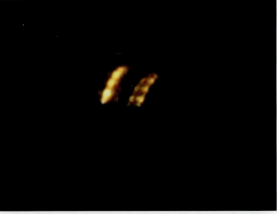

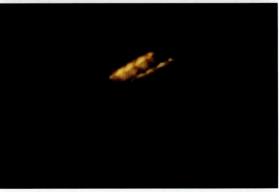

Horton H XVIII 1945

Left: The Horton H XVIII as developed and manufactured by the Germans at the end of WWII. Below: Idaho buisnessman Kenneth Arnold holds up a drawing of one of the craft that he saw from his small plane over Washington state in April of 1947. His drawing looks very much like the Horton H XVIII flying wing.

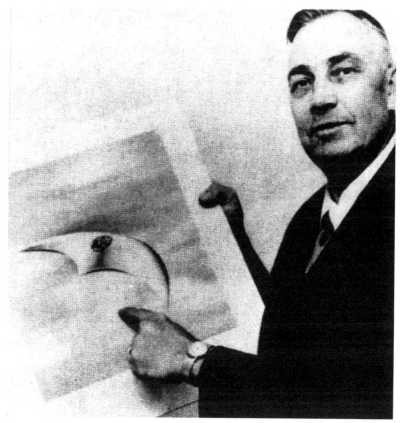

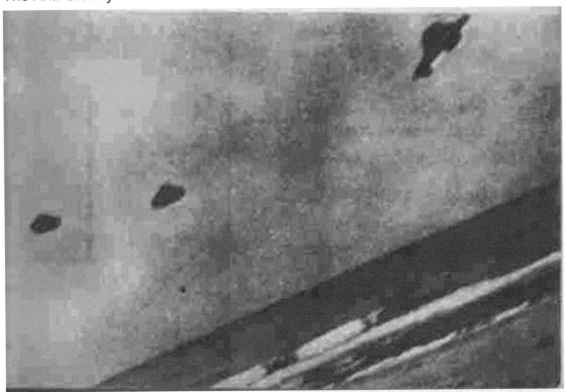

Two photos taken by Italian journalist Bruno Ghibaudi at the beach resort of Pescara, Italy in June of 1961. The second photo on the right seems to show wings or fluttering of the craft.

A photo taken in Rhode Island, USA, June 10, 1967.

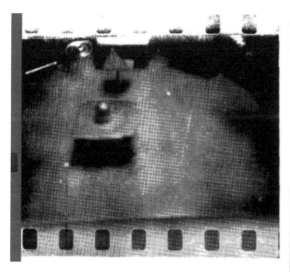

Two photos taken by Italian contactee Alberto Perego in 1958 of the interior of a craft that he said he was allowed to enter earlier that year.

The Cube!

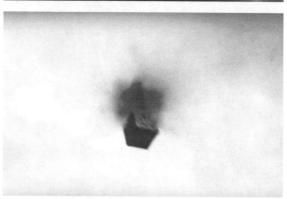

Two photos of a bizarre cube emerging from a cloud near El Paso, Texas on June 29, 2015.

99

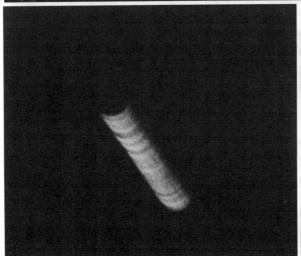

Above: Two amazing photos of the huge cylinder that hovered over New York City on March 20, 1950. *Right*: A British pilot named David Hastings took this photo of a cylindrical craft near the California-Nevada border, an area near Area 51.

A photo of "The Battle of Los Angeles" Feb. 25, 1942.

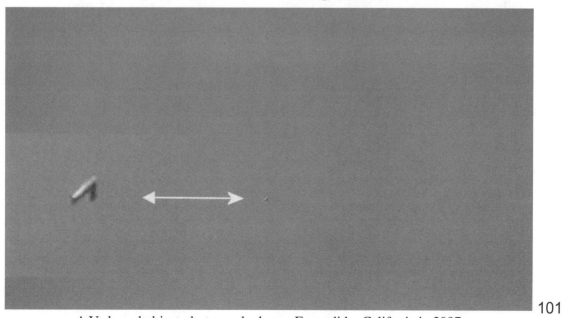

A V-shaped object photographed over Escondido, Califonia in 2007.

The Anti-Gravity Files

A series of three photos of the curious Ordzenikidze object located somewhere in Russia, provided in 1991 by the Russian cosmonaut Marina Popovich. A secret military project in Russia or a leftover from an unknown science fiction film?

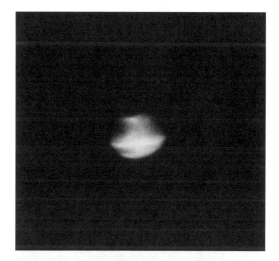

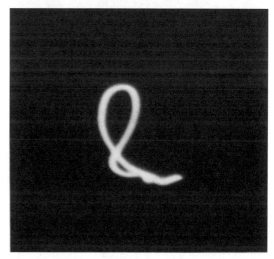

The Kaikoura UFO incident happened in New Zealand during December of 1978 when a large orb was seen flying near a cargo aircraft which began filming the glowing orb. In the third-to-last frame of the film the orb suddenly does a figure 8 maneuver for the camera, taking about 1/20 of a second.

An orb photographed over Hanover, PA in 1958

A photo of a triangular craft over Kiev, Ukraine in 1990.

A flying saucer photographed over Ticul, Yucatan, Mexico on Dec. 12, 2004.

Chapter 7
Death Rays
Anyone?

By

David Hatcher Childress

Archimedes of Syracuse using his death ray mirror, circa 212 BC. Painted by Guilio Parigi, 1599.

A Tesla tower in action in this 1920s illustration.

Everybody loves death rays. Death rays, death beams and rayguns were a theoretical weapon that gained popularity in the 1920s through the 1930s with depictions of the weapons in movies and serials like *The Invisible Ray, Buck Rogers* or *The Lost City*. They were either some kind of particle beam weapon or one that was electromagnetic in nature, giving a sudden burst, or pulse, of zapping electricity like that of a Tesla coil. Nikola Tesla was known to speak about both kinds of death rays to the press, often on his birthday. It has been claimed that various death ray devices were independently invented by Tesla, Guglielmo Marconi, Harry Grindell Matthews, Edwin R. Scott and others. Wikipedia notes that in 1957 the National Inventors Council was still mentioning a death ray device in its lists of needed military inventions.

The first inventor of a death ray may have been the Greek mathematician, physicist, engineer, inventor, and astronomer Archimedes of Syracuse. Born circa 287 BC, Archimedes of Syracuse is regarded as one of the great mathematicians and leading scientists of antiquity.

He anticipated modern calculus and analysis and invented all types of devices, including a death ray. One of Archimedes' more famous quotes is, "Give me a lever long enough and a place to stand and I will move the world."

During the two year siege of Syracuse (211 to 212 BC) Archimedes used catapults and heavy timbers to hurl objects at the Roman ships in the distance. He also created a giant iron claw that was operated from inside the city's walls, and outside the walls at the port, the claw was capable of picking up a Roman ship and plunging it back down into the water with the occupants

Boris Karloff with a death ray in *The Mask of Fu Manchu* (1932).

spilling into the water.

Archimedes' death ray was a series of mirrors that reflected concentrated sunlight onto Roman ships. The ships were moored within bow and arrow range to the city walls when the death ray burned into them with the collective, condensed sunlight beaming from these mirrors. The death ray was pointed at ship after ship, setting each on fire; they flaming ships had to be abandoned as they sunk into the Mediterranean. Archimedes was killed by Roman soldiers when they eventually stormed the city a few months later.

Tesla's Death Ray

Nikola Tesla was born on July 9, 1856 in Smiljan, a town in the Austrian Empire (now in Croatia). He died on January 7, 1943 in New York City. Tesla moved to the United States in 1884 and is best known as the inventor and engineer who discovered and patented the rotating magnetic field motor, the basis of most alternating-current machinery. In conjunction with his other inventions along these lines Tesla developed the three-phase system of electric power transmission. He invented the Tesla coil in 1891, an induction coil widely used in radio technology. He sold the patent rights to his system of alternating-current dynamos, motors, switches and transformers to George Westinghouse.

A Tesla-type death ray was featured in the 1932 movie *The Mask of Fu Manchu* which starred Boris Karloff as the original mad scientist and "Dr. Evil"—the diabolical Fu Manchu. Karloff appeared as a mad scientist with a death ray in the 1936 movie *The Invisible Ray*, itself a remake of a 1920 serial of the same name. Death rays had already been featured in many serials in the 1920s and Tesla was continually asked about them in the interviews he did with magazines and newspapers. On July 23, 1934 *Time* magazine wrote an article about Tesla's death ray:

> Last week Dr. Tesla announced a combination of four inventions which would make war unthinkable. Nucleus of the idea is a death ray—a concentrated beam of sub-microscopic particles flying at velocities approaching that of light. The beam, according to Tesla, would drop an army in its tracks, bring down squadrons of airplanes 250 miles away. Inventor Tesla would discharge the ray by means of 1) a device to nullify the impeding effect of the atmosphere on the particles; 2) a method for setting up high potential; 3) a process for amplifying that potential to 50,000.,000 volts; 4) creation of "a tremendous electrical repelling force.

Tesla claimed to have invented a "death beam" and a wall of energy that he called teleforce in the 1930s. Says Tesla in the 1937 article "A Machine to End War":

> Hitherto all devices that could be used for defense could also be utilized to serve for aggression. This nullified the value of the improvement for purposes of peace. But I was fortunate enough to evolve a new idea and to perfect means which

Nikola Tesla's interview in New York Times:
"Invisible Chinese wall of defense around USA" by Tesla's "teleforce."

The New York Times, July, 1934.

"Death Ray" for Planes

Nikola Tesla, one of the truly great inventors who celebrated his eighty-fourth birthday on July 10, tells the writer that he stands ready to divulge to the United States Government the secret of his "teleforce," with which, he said, airplane motors would be melted at a distance of 250 miles, so that an invisible Chinese Wall of Defense would be built around the country against any attempted attack by an enemy air force, no matter how large.

This "teleforce," he said, is based on an entirely new principle of physics that "no one has ever dreamed about," different from the principle embodied in his inventions relating to the transmission of electrical power from a distance, for which he has received a number of basic patents. This new type of force, Mr. Tesla said, would operate through a beam one one-hundred-dred-millionth of a square centimeter in diameter, and could be generated from a special plant that would cost no more than $2,000,000 and would take only about three months to construct.

A dozen such plants, located at strategic points along the coast, according to Mr. Tesla, would be enough to defend the country against all possible aerial attack. The beam would melt any engine, whether Diesel or gasoline-driven, and would also ignite the explosives aboard any bomber. No possible defense against it could be devised, he asserts, as the beam would be all-penetrating.

High Vacuum Eliminated

The beam, he states, involves four new inventions, two of which already have been tested. One of these is a method and apparatus for producing rays and other manifestations of energy in free air, eliminating the necessity for a high vacuum; a second one is a method and process for producing very great electrical force; a third is a method for amplifying this force, and a fourth is a new method for producing a tremendous repelling electrical force. This would be the projector, or gun, as it were, of the system. The voltage for propelling the beam to its objective, according to the inventor, will attain a potential of 50,000,000 volts.

With this enormous voltage, he said, microscopic electrical particles of matter will be catapulted on their mission of defensive destruction. He has been working on this invention, he added, for many years and has recently made a number of improvements in it.

Mr. Tesla makes one important stipulation. Should the government decide to take up his offer he would go to work at once, but they would have to trust him. He would suffer "no interference from experts."

In ordinary times such a condition would very likely interpose an insuperable obstacle. But times being what they are, and with the nation getting ready to spend billions for national defense, at the same time taking in consideration the reputation of Mr. Tesla as an inventor who always was many years ahead of his time, the question arises whether it may not be advisable to take Mr. Tesla at his word and commission him to go ahead with the construction of his teleforce plant.

Such a Device "Invaluable"

After all, $2,000,000 would be relatively a very small sum compared with what is at stake. If Mr. Tesla really fulfills his promise the result achieved would be truly staggering. Not only would it save billions now planned for air defense, by making the country absolutely impregnable against any air attack, but it would also save many more billions in property that would otherwise be surely destroyed no matter how strong the defenses are as witness current events in England.

Take, for example, the Panama Canal. No matter how strong the defenses, a suicide squadron of dive bombers, according to some experts, might succeed in getting through and cause such damage that would make the Canal unusable, in which case our Navy might find itself bottled up.

Considering the probabilities in the case even if the chances were 100,000 to 1 against Mr. Tesla the odds would still be largely in favor of taking a chance on spending $2,000,000. In the opinion of the writer, who has known Mr. Tesla for many years and can testify that he still retains full intellectual vigor, the authorities in charge of building the national defense should at once look into the matter. The sum is insignificant compared with the magnitude of the stake.

can be used chiefly for defense. If it is adopted, it will revolutionize the relations between nations. It will make any country, large or small, impregnable against armies, airplanes, and other means for attack. My invention requires a large plant, but once it is established it will be possible to destroy anything, men or machines, approaching within a radius of 200 miles. It will, so to speak, provide a wall of power offering an insuperable obstacle against any effective aggression.

If no country can be attacked successfully, there can be no purpose in war. My discovery ends the menace of airplanes or submarines, but it insures the supremacy of the battleship, because battleships may be provided with some of the required equipment. There might still be war at sea, but no warship could successfully attack the shoreline, as the coast equipment will be superior to the armament of any battleship.

I want to state explicitly that this invention of mine does not contemplate the use of any so-called " death rays." Rays are not applicable because they cannot be produced in requisite quantities and diminish rapidly in intensity with distance. All the energy of New York City (approximately two million horsepower) transformed into rays and projected twenty miles, could not kill a human being, because, according to a well known law of physics, it would disperse to such an extent as to be ineffectual.

My apparatus projects particles which may be relatively large or of microscopic dimensions, enabling us to convey to a small area at a great distance trillions of times more energy than is possible with rays of any kind. Many thousands of horsepower can thus be transmitted by a stream thinner than a hair, so that nothing can resist. This wonderful feature will make it possible, among other things, to achieve undreamed-of results in television, for there will be almost no limit to the intensity of illumination, the size of the picture, or distance of projection.

I do not say that there may not be several destructive wars before the world accepts my gift. I may not live to see its acceptance. But I am convinced that a century from now every nation will render itself immune from attack by my device or by a device based upon a similar principle.

At present we suffer from the derangement of our civilization because we have not yet completely adjusted ourselves to the machine age. The solution of our problems does not lie in destroying but in mastering the machine.

In the 2013 book from Princeton University Press, *Tesla: Inventor of the Electrical Age,* author W. Bernard Carlson says that Tesla had several different concepts for death rays including one that used mercury as a particle beam. Carlson shows a diagram of the nozzle that sprays the energized mercury into a death ray. He says that Tesla's plan was to accelerate tiny mercury particles to a velocity of 48 times the speed of sound.

To energize these particles Carlson says that Tesla proposed an electrostatic generator similar to the Robert Van de Graaff design for a Van de Graaff generator but in place of a charge-carrying belt he would use a circulating stream of desiccated air propelled by a Tesla pump—or blower—through hermetically sealed ductwork. This stream of air would pass two discharge points where it would be ionized by high-voltage direct current. The ions would then be carried up by the airstream where the charge would accumulate in a large spherical terminal similar to the one used on the top of Tesla's famous Wardenclyffe tower. To

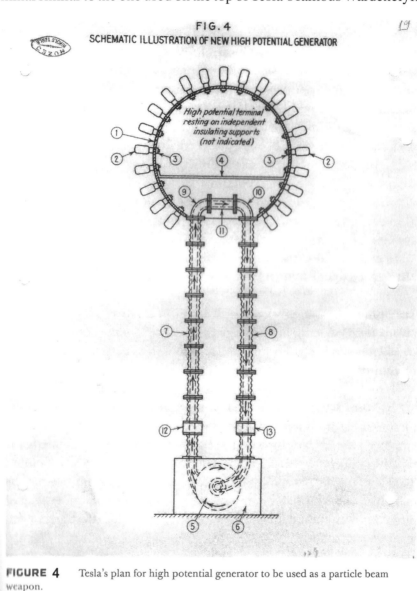

FIG. 4

SCHEMATIC ILLUSTRATION OF NEW HIGH POTENTIAL GENERATOR

FIGURE 4 Tesla's plan for high potential generator to be used as a particle beam weapon.

Key:
5	Tesla turbine or blower
7 and 8	ducts for air
12 and 13	points where air flow would be ionized by high-voltage direct current
1	sphere charged to high voltage
2	evacuated glass bulbs to increase sphere's electrical capacity

increase the electrical capacity of the spherical terminal, it was studded with evacuated glass bulbs, each of which contained an umbrella-shaped electrode.

Carlson says that Tesla wrote:

> I am confident that as much as one hundred million volts will be reached with such a transmitter providing a tool of inestimable value for practical purposes as well as scientific research.

Carlson says that within the sphere a high vacuum was to be maintained into which millions of tiny mercury particles would be introduced. Tesla said that these particles would be charged at the same high voltage of the whole sphere and accelerated out of the sphere through a specially designed projector-nozzle. The projector would shoot a single row of highly charged particles that would deliver prodigious amounts of energy over a great distance. A fascinating concept for a death ray—one that involved that mysterious element mercury: a metal, a liquid, a conductor.

It would seem that Tesla's primary death ray was the mercury particle beam. Particle beams are:

…special sorts of electromagnetic waves, a special sort of light.

The white light or daily light is a mixture of different lengths of waves. White light is a mixture of many colors which can be separated. Red light has long waves whereas blue light has short waves. The waves making up a particle beam are quite different. Not only are all the waves the same length, but they are lined up so that the tops (peaks) of the waves coincide with each other.

Particle beams can be concentrated into a tiny point. They have tremendous energy. Particle beams can produce enough heat to turn a metal into a vapor! They are accurate cutting tools that can even cut diamond, the hardest substance known to man. Particle beams are powerful enough to cut through metal in military operations; a particle beam can be bounced off a target such as an enemy airplane or ship to determine its distance and speed. Particle

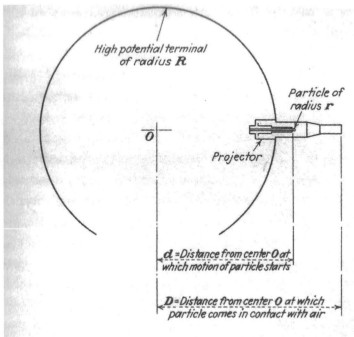

Diagram showing the projector that Tesla planned to use to shoot a stream of highly-charged mercury particles from his beam weapon.

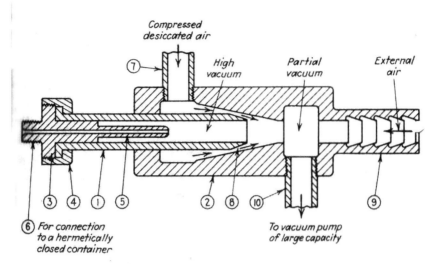

Diagram showing the projector nozzle for Tesla's mercury particle death ray.

gyroscopes (guidance devices) are being developed to direct bombs and artillery shells to their target. (Wikipedia, with changes to correct gramatical errors)

The search for death rays was a very real pursuit during WWII and the years leading up to the war. Militaries all over the world were looking for the latest in weapons—electronic and otherwise—that could be developed in the immediate future. If death rays were a real technology to be exploited by militaries—as movies and serials were showing on a daily basis and Nikola Tesla was hinting at in his press interviews—then governments around the world wanted to be on top of the subject. Says Wikipedia:

In the year 1923, Edwin R. Scott, an inventor from San Francisco, claimed he was the first to develop a death ray that would destroy human life and bring down planes at a distance. He was born in Detroit, and he claimed he worked for nine years as a student and protégé of Charles P. Steinmetz. Harry Grindell-Matthews tried to sell what he reported to be a death ray to the British Air Ministry in 1924. He was never able to show a functioning model or demonstrate it to the military.

…Antonio Longoria in 1934 claimed to have a death ray that could kill pigeons from four miles away and could kill a mouse enclosed in a "thick walled metal chamber."

During World War II, the Germans had at least two projects, and the Japanese one, to create so called death rays. One German project led by a man called Schiebold concerned a particle accelerator with a steerable bundle of beryllium rods running through the vertical axis. The other was developed by Dr. Rolf Wideroe and is referred to in his biography. The machine developed by Wideroe was in the Dresden Plasma Physics laboratory in February 1945 when the city was bombed.

Wideroe led a team in March 1945 to remove the device from the ruined laboratory and deliver it to General Patton's 3rd Army at Burggrub where it was taken into US custody on 14 April 1945. The Japanese weapon was called death ray "Ku-Go" which aimed to employ microwaves created in a large magnetron.

Most articles on particle beam weapons do not mention Tesla's version of charged mercury particles as being one of the means to create a particle beam death ray weapon. Perhaps it is part of the ongoing cover-up by the military industrial establishment. The main research facility at "Area 51" is known as Mercury. Wikipedia says:

A particle-beam weapon uses a high-energy beam of atomic or subatomic particles to damage the target by disrupting its atomic and/or molecular structure. A particle-beam weapon is a type of directed-energy weapon, which directs energy in a particular and focused direction using particles with negligible mass. Some particle-beam weapons are real and have potential practical applications, e.g. as an antiballistic missile defense system for the United States and its Strategic Defense Initiative. The vast majority, however, are science fiction and are among the most common weapon types of the genre. They have been known by myriad names: phasers, particle accelerator guns, ion cannons, proton beams, lightning rays,

Bela Lugosi and Boris Karloff face off with their death ray in *The Invisible Ray* (1936).

rayguns, etc. The concept of particle-beam weapons comes from sound scientific principles and experiments are currently underway around the world. One effective process to cause damage to or destroy a target is to simply overheat it until it is no longer operational.

Particle accelerators are a well-developed technology used in scientific research for decades. They use electromagnetic fields to accelerate and direct charged particles along a predetermined path, and electrostatic "lenses" to focus these streams for collisions. The cathode ray tube in many twentieth-century televisions and computer monitors is a very simple type of particle accelerator. More powerful versions include synchrotrons and cyclotrons used in nuclear research. A particle-beam weapon is a weaponized version of this technology. It accelerates charged particles (in most cases electrons, positrons, protons, or ionized atoms, but very advanced versions can use other particles such as mercury nuclei) to near-light speed and then shoots them at a target. These particles have tremendous kinetic energy which they impart to matter in the target's surface, inducing near-instantaneous and catastrophic superheating.

There have been officially acknowledged tests of military particle beam death rays. As part of Ronald Reagan's "Star Wars" program, the US Strategic Defense Initiative put into development at Alamos National Laboratory the technology of a neutral particle beam to be used as a weapon in outer space. As part of the Beam Experiments Aboard Rocket (BEAR) project, in July 1989 a prototype hydrogen beam weapon was launched from White Sands Missile Range that successfully deployed into low Earth orbit.

The prototype weapon was operated successfully in space and after reentry was recovered intact. It was then transferred from Los Alamos to the Smithsonian Air and Space Museum in Washington, DC in 2006. Naturally, no word of secret death ray-particle beam weapons and their testing has come from any other branch of the military nor from foreign governments such as the Chinese or Russians. Death ray technology is still considered highly top secret.

Wikipedia tells us that neutral particle beams are from ionized atoms:

Charged particle beams diverge rapidly due to mutual repulsion, so neutral particle beams are more commonly proposed. A neutral-particle-beam weapon ionizes atoms by either stripping an electron off of each atom, or by allowing each atom to capture an extra electron. The charged particles are then accelerated, and neutralized again by adding or removing electrons afterwards.

Cyclotron particle accelerators, linear particle accelerators, and Synchrotron particle accelerators can accelerate positively charged hydrogen ions until their velocity approaches the speed of light, and each individual ion has a kinetic energy range of 100 MeV to 1000 MeV or more. Then the resulting high energy protons can capture electrons from electron emitter electrodes, and be thus electrically neutralized. This creates an electrically neutral beam of high energy hydrogen atoms, that can proceed in a straight line at near the speed of light to smash into its target and damage it. The pulsed particle beam emitted by such a weapon may contain 1 gigajoule of kinetic energy or more. The speed of a beam approaching that of light (299,792,458 m/s in a vacuum) in combination with the energy created by the weapon would negate any realistic means of defending a target against the beam. Target hardening through shielding or materials selection would be impractical or ineffective, especially if the beam could be maintained at full power and precisely focused on the target.

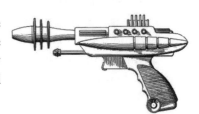

We are getting the sense that death rays and particle beam weapons are something that are beyond just science fiction from Hollywood and something that could easily exist—should exist—today. But where are the death rays and rayguns? Is there a big cover-up going on?

The Romance of Rayguns

Once we have our death ray technology all picked out we can move on to miniaturizing these high energy devices into practical rayguns for handy use when flying around the solar system trying to keep peace and order. Perhaps these energy directed pistols would use tiny charged particles of mercury, as Tesla imagined, or more conventional xenon and neon plasmas.

The miniaturization of these high-energy devices has been a major challenge since their inception and batteries that can store a large charge—or several large charges—have only recently been coming on to the market, as we have seen with the exploding batteries of certain brands of smartphones.

What little knowledge we have about rayguns largely comes from popular fiction starting with Buck Rogers and his Solar Scouts in the late 1920s. While it has not been proven that early versions of rayguns were produced in Europe, Japan and America, it has been speculated that small handheld rayguns were available starting in the 1930s, if not before.

In movies such as *The Invisible Ray* or *The Mask of Fu Manchu* the death ray is a large device and not one that could be put in a holster on one's side. Perhaps a supply of mercury and the sudden discharge of a 9-volt battery caused the kind of particle beam that could burn a hole through someone's head or heart in an atomic instant.

Rayguns as described by science fiction

The 1935 Marconi raygun, now a collector's item.

116

do not have the disadvantages that have, so far, made directed-energy weapons largely impractical as weapons in real life, calling for a suspension of disbelief by a technologically educated audience.

Ray guns draw seemingly limitless power from often unspecified sources. In contrast to their eal-world counterparts, the batteries or power packs of even handheld weapons are minute, durable, and do not seem to need frequent recharging.

Ray guns in movies are often shown as shooting discrete pulses of energy or a vortex-cyclonic energy beam that causes the victim to freeze or dematerialize. As to the many uses

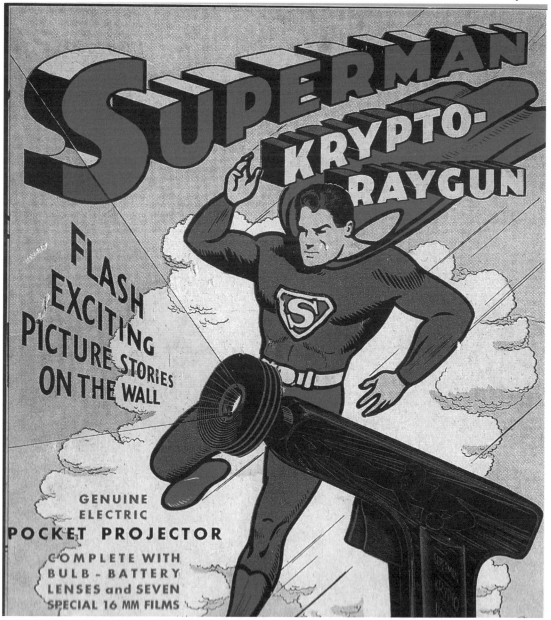

The Superman Krypto Raygun was all the rage in 1940.

The Anti-Gravity Files

for rayguns, Wikipedia says:

A wide range of non-lethal functions as determined by the requirements of the story: for instance, they may stun, paralyze or knock down a target, much like modern electroshock weapons. Occasionally the rays may have other effects, such as the "freeze rays" in the TV series *Batman* (1966–1968) and *Underdog* (1964–1970). Many of the more implausible functions are almost farcical and include rayguns that age or de-age people (various cartoons); shrink rays (*Fantastic Voyage, Honey, I Shrunk the Kids*), and a "dehydration ray" (*Megamind*).

In the end, rayguns are something that we are already living with, whether they are a reality or not. Our police today are equipped with stun guns that fire an electrical discharge from two electrodes. Hopefully, the rayguns of yore and death rays of Nikola Tesla mounted on the underside of a flying saucer will become public knowledge in the near future.

The Buck Rogers Rocket Pistol was only 50¢.

Chapter 8
The Tesla
Pyramid Engine

By
Ivar Slovac

The Anti-Gravity Files

THE TESLA PYRAMID ENGINE

Picture 1. Nikola Tesla

Nikola Tesla is probably the greatest inventor in the field of electricity and magnetism ever, one that will be hard to surpass. In the year 1977, on the occasion of the Celebration of the 120th Anniversary of the Birth of Nikola Tesla organized by the Yugoslav Academy of Arts and Science, Školska knjiga Zagreb company published Tesla's autobiography *My Inventions*.

In this book Tesla describes, among other things, his original idea and the purpose of constructing the Wardenclyff tower on Long Island, which never saw its completion.

One day, as I was roaming in the mountains, I sought shelter from an approaching storm. The sky became overhung with heavy clouds but somehow the rain was delayed until, all of a sudden, there was a lightning flash and a few moments after a deluge.

*This observation set me thinking. It was manifest that the two phenomena were closely related, as cause and effect, and a little reflection led me to the conclusion that the electrical energy involved in the precipitation of the water was inconsiderable, the function of lightning being much like that of a sensitive trigger. Here was a stupendous possibility of achievement. **If we could produce electric effects of the required quality, this whole planet and the conditions of existence on it could be transformed.** The sun raises the water of the oceans and winds drive it to distant regions where it remains in a state of most delicate balance. If it were in our power to upset it when and wherever desired, this mighty life-sustaining stream could be at will controlled. We could irrigate arid deserts, create lakes and rivers and provide motive power in unlimited amounts. This would be the most efficient way of harnessing the sun to the uses of man. The consummation depended on our ability to develop electric forces of the order of those in nature. It seemed a hopeless undertaking, but I made up my mind to try it and immediately on my return to the United States, in the summer of 1892, work was begun which was to me all the more attractive, because a means of the same kind was necessary for the successful transmission of energy without wires.*

TESLA'S RAIN ENGINEERING

Nikola Tesla saw clearly that water is life, the essential element in the origin and the development of life on the planet Earth. In arid regions there are hardly any life forms. Only water can make the desert green. Tesla wanted to make a device that would create lightning and thus cause rain, which would create a favourable climate in the desert.

Lightning is a phenomenon of electric discharge, and its audible manifestation is thunder. The entire phenomenon is called thunder. Even today, scientists are busy trying to understand and explain certain phenomena in relation with the thunder-strike. Condensation processes in the atmosphere cause accumulation of electricity in clouds. The polarisation of charges within storm-clouds leads to an increase in electric potential between certain parts, which results in electric discharge. Electric discharges can appear within the cloud, between two clouds or between the cloud and the ground. The latter is of interest to us, a giant short circuit between the sky and the ground.

It was as early as 1750 that Benjamin Franklin discovered that electricity is not evenly distributed over all elements of the surface of conductors (all except the sphere), which depends on the extent of curvature of the observed element of surface. Generally speaking, higher charge density is located on their edges, protuberant and pointed parts than on round and flat ones. Statistically, thunder always strikes along the path which is more conductive, that is to say the path with a higher charge density. Franklin's experiments found that the pointier and more conductive the lightning rod, the easier it was to attract lightning during a strike. That is why the lightning rod point is made of a metal rod with a gilded tip.

Benjamin Franklin invented the lightning rod – the technical solution to attract natural lightning from clouds, but Nikola Tesla wanted to make a device that would create lightning in cloudless desert areas and thus induce rainfall!

In the discharge tunnel of a natural lightning the temperature is about 30000°C, the electric charge reaches 40 MV, and the electric current 110 kA. It is estimated that the energy of an average lightning is about 40MWh.

The Wardenclyffe tower on Long Island designed by Tesla was supposed to use a transformer to create high voltage and instead of natural lightning, produce high energy ionic smouldering discharge. Tesla envisioned an 8-sided pyramid with a semi-sphere on top as a basis for this device. Why?

Picture 2. Wardenclyffe tower erected in Shoreham, Long Island in the state of New York. Tesla planned to use this tower as his world-range radio station.

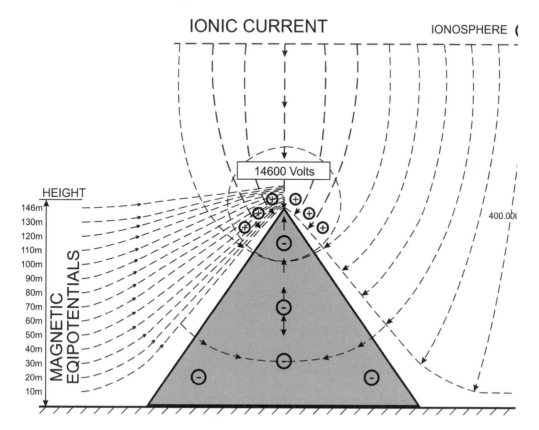

Picture 3. The effect of Cheops pyramid on the density of electric and magnetic fields Earth - ionosphere

Nikola Tesla claimed that Earth is a spherical capacitor plate with the ionosphere as the other plate. Recent measurements have determined that the voltage difference between the Earth and the ionosphere is 400.000 volts. The force lines of the electric charge plus fields that come from the Sun, act vertically onto the sides of the pyramids. Magnetic equipotentials show a great density of the magnetic field at the top. The electric field voltage grows 100V/m in height. The Earth's minus field is the greatest and at its most dense at the very top of the pyramid. At the top of Cheops pyramid there is voltage of 14600 volts. The Cheops pyramid has its own electric capacity – ability to accumulate a certain amount of electric charge. If too much electric charge is brought to the Cheops pyramid, the excess of these charges discharges at the top of the pyramid. According to the word of mouth, the top of the Cheops pyramid was originally a pyramidon made of solid gold – an excellent conductor.

The voltage at the top of the structure depends solely on height. That is why Tesla built a tower, a tall-standing structure. Tesla chose an 8-sided pyramid –and it could have been a four-sided one, or infinite-sided – a cone. The voltage would have been the same in all cases. The symmetrical form of the structure increases its static stability and resistance to earthquakes.

Why did Tesla build a sphere, 21 m in diameter on top of a pyramidal tower? What was accomplished by doing that? The fact that the sphere is spike-less, the voltage on the sphere is evenly distributed over its entire surface. The voltage that would have discharged from the top of the pyramid was taken over by the sphere, thus increasing the total electric capacity of the whole tower. Instead of having the discharge take place on one point on the top of the pyramid, now it takes place in many points all over the surface of the sphere.

Conclusion: by adding the sphere to the top of the pyramid, the electric capacity of the device was increased, and this increased the possibility to create far stronger lightning.

The Tesla tower was designed to electrify the atmosphere in a split second and produce thunder and rainfall. To our great misfortune, the Tesla tower on Long Island was never completed. It was dismantled during the First World War, and Tesla never got the chance to experiment with artificial thunder and rain engineering in desert conditions.

NEGATIVE IONS

How do state-of-the-art air-conditioning devices work?
They regulate air temperature and humidity and generate negative ions.

The influence of negative ions to human health has been known for about 60 years now, and there has been a great deal of scientific research. There are constant bio-chemical processes of oxidation and reduction taking place in the human body, which require negatively charged ions to function properly. All living things receive negatively charged ions from the air through the skin and respiration organs. When there are enough negative ions in a room, there is an increase in oxygen content in the blood, heart rate gets normalized and this accelerates the excretion of toxins. Due to better blood-flow in the brain, concentration and other mental capabilities strengthen, the consequences of stress get neutralized, tension and headaches are relieved. Negative ions bind with toxins in the body, which means they act as anti-oxidants, thus preventing the development of degenerative disease.

In nature the most favourable condition of air ionization is after thunderstorms, cloudburst and downpours, when atmospheric discharges occur. That is when there are noticeably more negative ions in the air than there are positive ones. Increased concentrations of negative oxygen ions are also found near waterfalls, in pine forests, mountains, along the sea shore and in caves, which led to speleotherapy (http://www.showcaves.com/english/explain/Misc/Speleotherapy.html).

PYRAMIDS – GENERATORS OF NEGATIVE IONS

What was the function of pyramids? To produce thunder? No, it wasn't. For thousands of years after having been built, pyramids have functioned and continue to function as generators of negative ions. In order for pyramids to continuously ionize the surrounding air, they needed to be connected to a permanent source of negative ions. How was this done? Where are these energy power lines which lead to the pyramids?

The Giza Plateau hides an abundance of underground water. Huge underground rivers flowing around the pyramids are full of negative ions and by piezoelectric effect they transpond them to the pyramid, which then accumulates them and the surplus discharges on the top. All the great pyramids were built from stone with high content of crystal capable of binding electric charges from water when underground water mechanically presses against the crystal. If we expose the crystal to mechanical deformation, bound electric charges appear on its surface. This phenomenon is called piezoelectric effect. The crystal is the converter of mechanical into electric energy. Due to continuous charging and discharging off the pyramid electrically charged from underground rivers, the pyramid contracts and expands, which results in continuous micro-quakes.

OBELISKS, CHURCHES AND MOSQUES – IONIZERS

Obelisks function on totally identical principle. They are pillars made of crystalline stone, located over underground water, with a pyramidon made of gold, silver or copper on top of it. An obelisk is actually a miniature air ionizer.

The above mentioned principle has been applied on old churches (mosques) which were built using stone with high crystalline content, where the church tower was mostly made of copper, silver or gold. Almost all old churches (mosques) were located over underground water flows, a natural source of negative ions. A favourable negative ionization can be sensed even inside the church (mosque) itself.

STONE MEGALITHS – IONIZERS

Standing stone megaliths which we find all over the world are the most primitive form of air ionizers. They are made of crystal material and they are located above the underground water flows, but they do not have the pointed top made of conductive material and therefore they have less power of air ionization. This power deficiency was solved by making larger number of megaliths in some area, which increased the total power.

Bosnian standing stones also belong to that group – unique monumental stone blocks which are densely spread over the territory of Bosnia and Herzegovina, southern parts of Croatia, western Montenegro and southwestern Serbia. Standing stone blocks are a common national custom and as tomb stones they are a mark of the medieval culture in the mentioned regions.

Picture 4 and 5 Bosnian standing stone

Picture 6. Positioning and density of Bosnian standing stones

Dubravko Lovrenović, Ph.D., a member of the Committee for Preservation of National Monuments of Bosnia and Herzegovina, estimates that in Bosnia and Herzegovina alone there are over 100,000 Bosnian standing stones.

Although Bosnian standing stones are as a rule scattered and relocated, and barrows have been rebuilt many times, they are found in streaks. Almost without exceptions, the directions of streaks are south – north. It is estimated that the real number of stone blocks is much bigger than the number of registered Bosnian standing stone. Many of them were lost, destroyed, used for construction or simply were not found. **Stone blocks - monuments**, fractured in certain time and space, are a heritage and autochthonous treasury of an old civilization which has remained unknown to archeologists and historians.

TUNNELS AND PYRAMIDS

While Bosnian standing stone and obelisk exclusively have an outward effect, favorable ionization can be felt within the hollow spaces of church and pyramid.

Apart from underground water, the pyramid uses an additional source of negative ions – natural cave. The Cheops pyramid has a cave, and pyramids which do not have a cave are connected by underground tunnels with a remote one. Underground tunnels around the pyramids present an "artificial cave" and spending time in them is very good for one's health.

Results of radar measurements on the Giza plateau by Egyptian geophysicists Abbas Mohamed Abbas, El-said A. El-Sayed, Fathy A. Shaaban and Tarek Abdel-Hafez were published in the Egyptian national geophysics magazine, special edition, PP 1-16, (2006), Learn more on: http://www.howtosurvive2012.com/pdf_files/Pyramids.pdf

350 m southeast of the Cheops pyramid, they registered a signal of a potential tunnel, with dielectric values of underground structures which are presented in picture 7.

The electric capacity of a conductor which is located in the isolator of dielectric constant ε equals: $C = \varepsilon\, C_0$, where C_0 presents its capacity in vacuum.

The larger dielectric constant "ε" of the material, the larger the capacity of the capacitor. For vacuum $\varepsilon = 1$, hydrogen $\varepsilon = 1.2$, oxygen $\varepsilon = 1.5$, silicone $\varepsilon = 4.5$, limestone $\varepsilon = 4\text{-}8$, water $\varepsilon = 81$. The more moisture there is in the tunnel ($\varepsilon 1$), the more electric charges the air can accumulate. In order to prevent the limestone from taking all these negative charges from the Giza plateau ($\varepsilon 2$), the original constructors of the pyramid complex in Giza "isolated" the tunnel with stones ($\varepsilon 3$) built into the walls of the tunnel.

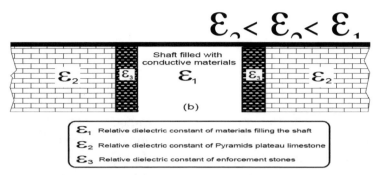

ε_1 – relative dielectric constant of materials filling the shaft
ε_2 – relative dielectric constant of the pyramids plateau limestone
ε_3 – relative dielectric constant of enforcement stones

Picture 7. The figure shows the possible presence of shafts in the studied areas

EGYPTIAN AND BOSNIAN PYRAMIDS – GIANT OSCILLATORS

Russian geophysicists O. B. Khavroshkin, Ph.D. and V.V. Tsyplakov, Ph.D. from the Moscow Schmidt Institute of Physics of the Earth have been analyzing micro-quakes on pyramids for many years. They have conducted measurements of seismic noises on about sixty Egyptian pyramids, and I personally met them in August of 2007 when they were doing measurements on Bosnian pyramids.

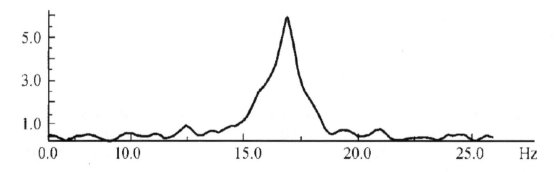

Picture 8. Spectrum of seismic noise from the foot of the south side of curved (Snofru) pyramid in Dahkshur region, Egypt

Many theories were developed on the reasons behind micro-quakes on pyramids. At the foot of the south side of Snofru's pyramid, they registered stronger harmonics of the micro-quake of 17 Hz. 17 Hz belongs in the infrasound range (0 Hz – 20 Hz) which cannot be heard by human ear. How can we explain the cause of this continuous micro-quake? Asuan hydro power plant is about 40 km away and it is hard to assume that the registered sub-harmonics (f_0=50Hz/3=16.67Hz) comes from electrical generators. Is it possible that Snofru's pyramid is actually a radio station receiving a signal from pulsar PSR 1913+16 which has the frequency of 16.95 Hz?

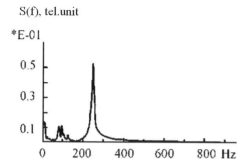

Picture 9. Spectrum of seismic noise from the Red pyramid (Dahkshur)

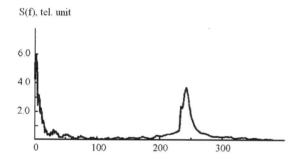

Picture 10. Spectrum of seismic noise from Menkaure's pyramid

We can see that Egyptian pyramids are musical instruments tuned to different tone frequencies. In Bosnian pyramids we encounter the same phenomenon.

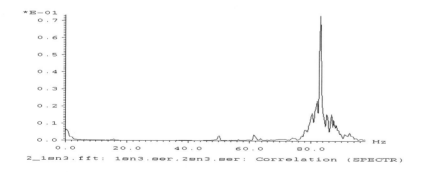

Picture 11. Spectrum of seismic noise from the foot of the Bosnian Pyramid of the Sun, f_0=81Hz

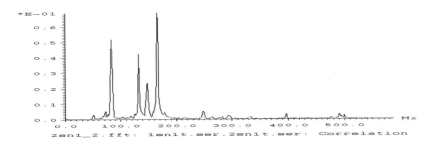

Picture 12. Spectrum of seismic noise from the top of the Bosnian Pyramid of the Sun, f=83Hz, 132Hz, 147Hz and 165Hz

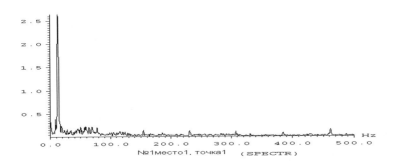

Picture 13. Spectrum of seismic noise at the foot of the Temple of the Earth, f=13Hz

AUTOMATIC REGULATION OF PYRAMIDS OPERATION

From the appended charts of the Russian geophysicists, we can see that every pyramid shaped structure has different frequency characteristics. It is completely logical, because there are no two pyramids of the same dimensions, construction material, or underground structure. It is clear that every pyramid concentrates seismic waves using one or more resonant frequencies. Depending on atmospheric electricity, the pyramid changes the amount of negative electric charges which it draws from underground waters and tunnels, releasing them into the atmosphere, and due to the piezoelectric effect, the amplitudes and frequencies of micro-quakes are slightly changed as well.

Electric current will start to flow between two differently charged elements, the power of which depends on the difference of their potential. The larger the voltage, the stronger the current. The lower the voltage, the weaker the current. The sunnier it is, the air is richer in positive charges and then the pyramid draws more negative ions from the underground

130

waters and tunnels in order to neutralize the positive charges in the atmosphere. After the rain, the air in the atmosphere is poorer in positive ions and then the pyramid loses the power of generating negative ions.

SUMMARY

A crystalline pyramid, with the belonging underground rivers and tunnels, functions as a generator of useful negative ions with automatic power regulation.

In 1892 Nikola Tesla got the idea to create a machine for the production of rain which would create favorable conditions for developing life in some regions. This machine of Tesla's was supposed to work on the principle of strong ionization of atmosphere in a split second.

Instead of Tesla's shocking ionization of atmosphere, the original pyramid builders were in possession of a technology capable of less powerful, but continuous and almost unnoticeable ionization of atmosphere.

Thousands of years after having been built, Bosnian pyramids still function excellently, to this day. Bosnian pyramids are covered in green vegetation and the landscape around them is marked with green valleys with plenty of benefitial negative electric charges which come from under the pyramids. Continuous ionization of the atmosphere preserves local eco systems. The whole Bosnian Valley of pyramids is a network of underground tunnels and underground rivers, which is confirmed by numerous ground-water aquifiers.

The desert landscape with no green vegetation around Egyptian pyramids brings me to the conclusion that they are not functioning well. There must have been some kind of failure in the system. Perhaps the underground tunnels caved in. Perhaps somebody deliberately closed some of the tunnels and therefore obstructed the stream of negative ions. Perhaps some underground waters have dried out or perhaps a great flood caused redirection of underground water flows. It remains a mistery.

Considering all of the above, we can see that the original pyramid builders were technologically more progressive than Nikola Tesla. That is why it is hard to believe the theory of Egyptian pharaohs organizing the construction of Egyptian pyramids – tombs. Egyptian pyramids were built long before the first appearance of dynasties in Egypt, by a civilization unknown to us, which has not yet been discovered by archaeology or history.

Reprinted from the September, 1927
issue of Science and Invention.

Gravity Nullified

Quartz Crystals Charged by High Frequency Currents Lose Their Weight

ALTHOUGH some remarkable achievements have been made with short-wave low power transmitters, radio experts and amateurs have recently decided that short-wave transmission had reached its ultimate and that no vital improvement would be made in this line. A short time ago, however, two young European experimenters working with ultra short-waves, have made a discovery that promises to be of primary importance to the scientific world.

The discovery was made about six weeks ago in a newly established central laboratory of the Nessartsaddin-Werke in Darredein, Poland, by Dr. Kowsky and Engineer Frost. While experimenting with the constants of very short waves, carried on by means of quartz resonators, a piece of quartz which was used, suddenly showed a clearly altered appearance. It was easily seen that in the center of the crystal, especially when a constant temperature not exceeding ten degrees C. (50 degrees Fahrenheit) was maintained, milky cloudiness appeared which gradually developed to complete opacity. The experiments of Dr. Meissner, of the Telefunken Co., along similar lines, according to which quartz crystals, subjected to high frequency currents clearly showed air currents which led to the construction of a little motor based on this principle. A week of eager experimenting finally led Dr. Kowsky and Engineer Frost to the explanation of the phenomenon, and further experiments showed the unexpected possibilities for technical uses of the discovery.

Some statements must precede the explanation. It is known at least in part, that quartz and some other crystals of similar atomic nature, have the property when exposed to potential excitation in a definite direction, of stretching and contracting; and if one uses rapidly changing potentials, the crystals will change the electric waves into mechanical oscillations. This *piezo electric* effect, shown in Rochelle salt crystals by which they may be made into sound-producing devices such as loud speakers, or reversely into microphones, also shows the results in this direction. This effect was clearly explained in August, 1925 *Radio News* and December, 1919 *Electrical Experimenter*. These oscillations are extremely small, but have nevertheless their technical use in a quartz crystal wave-meter and in maintaining

Fig. 1. The gravitation mullifier is shown in this illustration. The quartz crystal may be seen supporting a 55-pound weight. Dr. Kowsky is shown in a top coat because of the temperature at which the experiments were performed.

a constant wavelength in radio transmitters. By a special arrangement of the excitation of the crystal in various directions, it may be made to stretch or increase in length and

Fig. 3. This shows how the quartz crystal lost weight when subjected to the high frequency current. The original crystal was balanced on the scale.

will not return to its original size. It seems as if a dispersal of electrons from a molecule resulted, which, as it is irreversible, changes the entire structure of the crystal, so that it cannot be restored to its former condition.

The stretching out, as we may term this strange property of the crystal, explains the impairment of its transparency. At the same time a change takes place in its specific

gravity. Testing it on the balance showed that after connecting the crystal to the high tension current, the arm of the balance on which the crystal with the electrical connections rests, rose into the air. The illustration, Fig. 3, shows this experiment.

This pointed the way for further investigation and the determination of how far the reduction of the specific gravity could be carried out. By the use of greater power, finally to the extent of several kilowatts and longer exposure to the action, it was found eventually that from a little crystal, 5 by 2 by 1.5 millimeters, a non-transparent white body measuring about ten centimeters on the side resulted, or increased about 20 times in length on any side (see Fig. 4.) The transformed crystal was so light that it carried the whole apparatus with itself upwards, along with the weight of twenty-five kilograms (55 lbs.) suspended from it and floating free in the air. On exact measurement and calculation, which on account of the excellent apparatus in the Darredein laboratory could be readily carried out, it was found that the specific gravity was reduced to a greater amount than the change in volume would indicate. Its weight had become practically negative.

There can be no doubt that a beginning has been made toward overcoming gravitation. It is to be noted, however, that the law of conservation of energy is absolutely unchanged. The energy employed in treating the crystal, appears as the counter effect of gravitation. Thus the riddle of gravitation is not fully solved as yet, and the progress of experiments will be followed further. It is, however, the first time that experimentation with gravitation, which hitherto has been beyond the pale of all such research, has become possible, and it seems as if there were a way discovered at last to explain the inter-relations of gravity with electric and magnetic forces, which connection, long sought for, has never been demonstrated. This report appears in a reliable German journal, "Radio Umschau."

Fig. 2. The schematic diagram of the experiment is shown in this illustration. The high frequency oscillator has been omitted for clearness.

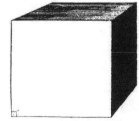

Fig. 4. This illustration shows the relative sizes of the crystal before and after the experiment. It is approximately twenty times its original length on any side.

Don't fail to see our next issue regarding this marvelous invention.

Chapter 8
Mercury Gyros,
Foo Fighters and
Quantum Vacuum Thrusters

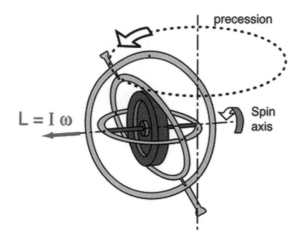

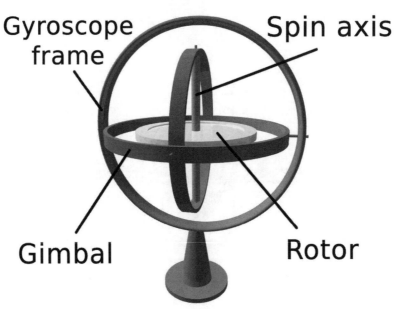

Gyroscope frame

Spin axis

Gimbal

Rotor

A standard gyroscope.

The topic of gyroscopes as a method of thrust is one that crops up from time to time though it is difficult to find very much material on the subject. The 1980 book *The Death of Rocketry* by Joel Dickenson and Robert Cook promoted Cook's invention known as the CIP device (Cook Inertial Propulsion) which was studied briefly by the Boeing company in Seattle.

Such devices are typically known as gyroscopic inertial thrusters, or GIT drives. Wikipedia has this brief paragraph on gyroscopic inertial thrusters:

> The Gyroscopic Inertial Thruster is a proposed reactionless drive based on the mechanical principles of a rotating mechanism. The concept involves various methods of leverage applied against the supports of a large gyroscope. The supposed operating principle of a GIT is a mass traveling around a circular trajectory at a variable speed. The high-speed part of the trajectory allegedly generates greater centrifugal force than the low, so that there is a greater thrust in one direction than the other. Scottish inventor Sandy Kidd, a former RAF radar technician, investigated the possibility (without success) in the 1980s. He posited that a gyroscope set at various angles could provide a lifting force, defying gravity. In the 1990s, several people sent suggestions to the Space Exploration Outreach Program (SEOP) at NASA recommending that NASA study a gyroscopic inertial drive, especially the developments attributed to the American inventor Robert Cook and the Canadian inventor Roy Thornson. In the 1990s and 2000s, enthusiasts attempted the building and testing of GIT machines. Eric Laithwaite, the "Father of Maglev," received a US patent for his "Propulsion System," which was claimed to create a linear thrust through gyroscopic and inertial forces. After years of theoretical analysis and laboratory testing of actual devices, no rotating (or any other) mechanical device has ever been found to produce unidirectional reactionless thrust in free space.

The Wikipedia article mentions the Laithwaite gyro which the Canadian author and scientist T.B. Pawlicki used in the plans for his flying saucer in the 1981 book *How to Build a Flying Saucer*. Pawlicki claimed that by mounting

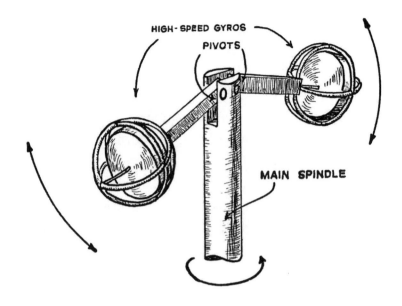

ESSENTIAL DESIGN
OF THE LAITHWAITE ENGINE

WHEN THE MAIN SPINDLE ROTATES,
PRECESSIONAL ACCELERATION
CAUSES THE GYROSCOPES TO RISE
AND FALL DURING EACH REVOLUTION.
THE MECHANICAL PROBLEM IS TO
ENGAGE THE RISE OF THE
GYROSCOPES TO GENERATE A LIFT
WHILE DISENGAGING THE
DOWNWARD SWING.

gyros around a craft a workable flying saucer could be built. Neither Pawlicki nor the Wikipedia article mention mercury gyros however. This seems to be a subject that has eluded most researchers. UFO researcher Bill Clendenon began mentioning mercury gyros at UFO conferences in the 1960s and said that the ancient vimanas of Hindu epics used mercury gyros and that the foo fighters seen over Germany at the end of WWII were mercury gyros as well.

Mercury, Clendenon pointed out, was an element, a metal, a liquid and a conductor. Its properties make it unique among the elements. The liquid metal mercury, when heated by any means, gives forth a hot vapor that is deadly. Mercury is generally confined to glass tubes or containers that are sealed, and therefore harmless to the user. High-frequency sound waves produce bubbles

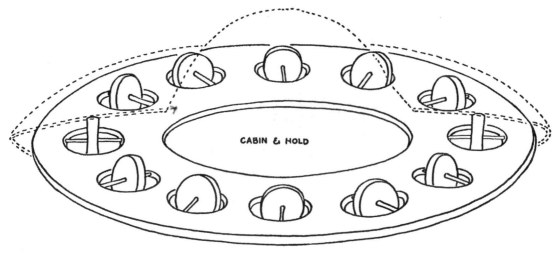

SCHEMATIC DIAGRAM
MARK I FLYING SAUCER

A SERIES OF ELECTRONIC CENTRIFUGES BASED ON THE HYPERSPACE OR PLANETARY GEAR DRIVES AND MOUNTED OUTBOARD ON A COWLED DISC ESTABLISHES THE CHARACTERISTIC PROFILE OF THE FLYING SAUCER.

From T.B. Pawlicki's book *How to Build a Flying Saucer*.

in liquid mercury. When the frequency of the bubbles grow to match that of the sound waves the bubbles implode, releasing a sudden burst of heat. Also, a circular dish of mercury revolves in a contrary manner to a naked flame circulated below it, and it gathers speed until it exceeds the speed of revolution of said flame.

Mercury was also the messenger god of the ancient Greeks and was known to fly through the sky. Clendenon pointed out that the symbol of the god mercury was the caduceus staff—a rod entwined by two serpents and topped with a winged sphere. Today the caduceus is used by the medical profession as its symbol, a practice that apparently stems from the Middle Ages. It was said that if the gods wanted to communicate, carry on commerce, or move things swiftly from one place to another over a long distance safely, they made use of Mercury to accomplish their goals.

Mercury wore winged sandals and a winged hat which bore him over land and sea with great speed. He carried with him his magic wand or "caduceus"— the winged staff with which he could perform many wondrous feats. In one

137

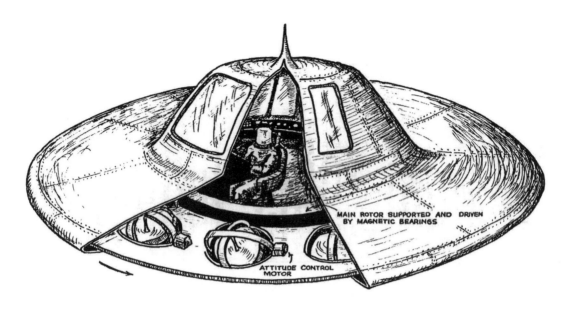

MARK II FLYING SAUCER

ELECTRONIC CENTRIFUGES BASED ON THE VORTEX DRIVE ARE MOUNTED IN GIMBALS TO TURN IN SYNCH WITH THE REVOLUTIONS OF THE MAIN ROTOR DISC.
THE TUNED ELECTROMAGNETIC FIELD GENERATED BY THE VORTEX DRIVE CAUSES THE VEHICLE TO BE CARRIED BY THE EARTH'S ELECTROMAGNETIC FIELD LIKE A DIRIGIBLE ELECTRON.
CONTROLLED GEOMAGNETIC PROPULSION IMPROVES THE DESIGN EFFICIENCY TO THE MARK III STAGE

From T.B. Pawlicki's book *How to Build a Flying Saucer*.

form or another, the ancient symbol has appeared throughout the ages.

Clendenon claimed that the caduceus was symbolic of the mercury gyros and of "electromagnetic flight and cosmic energy." The entwined snakes are

the vortex coils of the propellant, the rod the mercury boiler/starter/ antenna, and the wings symbolic of flight. He believed that the Germans had rediscovered this technology during WWII.

Since mercury can be a conductor of electricity it is possible to send an electric charge into the whirling mass of mercury within the gyro and electrify it into a charged gas known as a plasma.

The caduceus as a mercury engine.

This mercury plasma inside a sealed glass ball that is also a gyro will give an anti-gravity effect that is stronger than a standard gyroscopic effect, which is already fairly powerful. Just how many types of mercury-vortex-gyro devices their might be remains unknown.

Clendenon says that a mercury gyro propelled craft would be a discoid shaped flying saucer craft that works basically like this:

A gyroscope can be filled with mercury.

•The electromagnetic field coil that consists of the closed circuit heat exchanger/condenser coil circuit containing the liquid metal mercury and/or its hot vapor, is placed with its core axis vertical to the craft.

•A ring conductor (directional gyro-armature) is placed around the field coil (heat exchanger) windings so that the core of the vertical heat exchanger coils protrudes through the center of the ring conductor.

•When the electromagnet (heat exchanger coils) is energized, the ring conductor is instantly shot into the air, taking the craft as a complete unit along with it.

•If the current is controlled by a computerized resistance (rheostat), the ring conductor armature and craft can be made to hover or float in the Earth's atmosphere.

CONVENTIONAL MOTOR

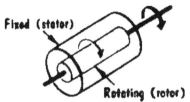

Fixed (stator)

Rotating (rotor)

GYRO MOTOR

Rotor

Stator

MOST MOTORS HAVE THE ROTATING PART ON THE INSIDE
AND THE FIXED PART ON THE OUTSIDE
BUT GYRO MOTORS ARE INSIDE OUT

•The electromagnet hums and the armature ring (or torus) becomes quite hot. In fact, if the electrical current is high enough, the ring will glow dull red or rust orange with heat.

•The phenomenon (outward sign of a working law of nature) is brought about by an induced current effect identical with an ordinary transformer.

•As the repulsion between the electromagnet and the ring conductor is mutual, one can imagine the craft being affected and responding to the repulsion phenomenon as a complete unit.

•Lift or repulsion is generated because of the close proximity of the field magnet to the ring conductor. Clendenon says that lift would always be vertically opposed to the gravitational pull of the planet Earth, but repulsion can be employed to cause fore and aft propulsion.

Foo Fighters

Clendenon, as well as a large number of other investigators, believe that the Germans rediscovered this mercury gyro technology during their research before and during WWII. He thinks that the Germans were also analyzing the Hindu texts on vimanas and successfully developed small flying mercury

The messenger god Mercury with his winged helmet and winged sandals.

gyros known as foo fighters that would robotically fly around the allied bombers and discharge an electromagnetic pulse that would interfere with the engines of the craft and make them turn back to their base.

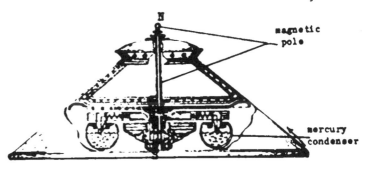

But first, let us take a quick look at the official view of foo fighters. The following is from Wikipedia:

The term foo fighter was used by Allied aircraft pilots in World War II to describe various UFOs or mysterious aerial phenomena seen in the skies over both the European and Pacific theaters of operations.

Though "foo fighter" initially described a type of UFO reported and named by the U.S. 415th Night Fighter Squadron, the term was also commonly used to mean any UFO sighting from that period. Formally reported from November 1944 onwards, witnesses often assumed that the foo fighters were secret weapons employed by the enemy. The Robertson Panel explored possible explanations, for instance that they were electrostatic phenomena similar to St. Elmo's fire, electromagnetic phenomena, or simply reflections of light from ice crystals.

Etymology

The nonsense word "foo" emerged in popular culture during the early 1930s, first being used by cartoonist Bill Holman, who peppered his Smokey Stover fireman cartoon strips with "foo" signs and puns. The term 'foo' was borrowed from Bill Holman's Smokey Stover by a radar operator in the 415th

The popular comic strip Smokey Stover.

Night Fighter Squadron, Donald J. Meiers, who (it is agreed by most 415th members) gave the foo fighters their name. Meiers was from Chicago and was an avid reader of Bill Holman's strip which was run daily in the *Chicago Tribune*. Smokey Stover's catch-phrase was "where there's foo, there's fire." In a mission debriefing on the evening of November 27, 1944, Fritz Ringwald, the unit's S-2 Intelligence Officer, stated that Meiers and Ed Schleuter had sighted a red ball of fire that appeared to chase them through a variety of high-speed maneuvers. Fritz said that Meiers was extremely agitated and had a copy of the comic strip tucked in his back pocket. He pulled it out and slammed it down on Fritz's desk and said, "…it was another one of those fuckin' foo fighters!" and stormed out of the debriefing room.

According to Fritz Ringwald, because of the lack of a better name, it stuck. And this was originally what the men of the 415th started calling these incidents: "Fuckin' Foo Fighters." In December 1944, a press correspondent from the Associated Press in Paris, Bob Wilson, was sent to the 415th at their base outside of Dijon, France to investigate this story. It was at this time that the term was cleaned up to just "foo fighters." The unit commander, Capt. Harold Augsperger, also decided to shorten the term to "foo fighters" in the unit's historical data.

History

The first sightings occurred in November 1944, when pilots flying over Germany by night reported seeing fast-moving round glowing objects following their aircraft. The objects were variously described as fiery, and glowing red, white, or orange. Some pilots described them as resembling Christmas tree lights and reported that they seemed to toy with the aircraft, making wild turns before simply vanishing. Pilots and aircrew reported that the objects flew formation with their aircraft and behaved as if under intelligent control, but never displayed hostile behavior. However, they could not be outmaneuvered or shot down. The phenomenon was so widespread that the lights earned a name—in the European Theater of Operations they were often called "kraut fireballs" but

An allied bomber painted with Smokey Stover.

for the most part called "foo-fighters." The military took the sightings seriously, suspecting that the mysterious sightings might be secret German weapons, but further investigation revealed that German and Japanese pilots had reported similar sightings.

On 13 December 1944, the Supreme Headquarters Allied Expeditionary Force in Paris issued a press release, which was featured in the *New York Times* the next day, officially describing the phenomenon as a "new German weapon." Follow-up stories, using the term "Foo Fighters," appeared in the *New York Herald Tribune* and the *British Daily Telegraph*.

In its 15 January 1945 edition, *Time* magazine carried a story entitled "Foo-Fighter," in which it reported that the "balls of fire" had been following USAAF night fighters for over a month, and that the pilots had named it the "foo-fighter." According to *Time*, descriptions of the phenomena varied, but the pilots agreed that the mysterious lights followed their aircraft closely at high speed. Some scientists at the time rationalized the sightings as an illusion probably caused by after-images of dazzle caused by flak bursts, while

A 1940 German torpedo gyro direction keeper.

others suggested St. Elmo's Fire as an explanation.

The "balls of fire" phenomenon reported from the Pacific Theater of Operations differed somewhat from the foo fighters reported from Europe; the "ball of fire" resembled a large burning sphere which "just hung in the sky," though it was reported to sometimes follow aircraft. On one occasion, the gunner of a B-29 aircraft managed to hit one with gunfire, causing it to break up into several large pieces which fell on buildings below and set them on fire. There was speculation that the

Foo fighters photgraphed around this fighter.

phenomena could be related to the Japanese fire balloons campaign. As with the European foo fighters, no aircraft were reported as having been attacked by a "ball of fire."

The postwar Robertson Panel cited foo fighter reports, noting that their behavior did not appear to be threatening, and mentioned possible explanations, for instance that they were electrostatic phenomena similar to St. Elmo's fire, electromagnetic phenomena, or simply reflections of light from ice crystals. The Panel's report suggested that "If the term 'flying saucers' had been popular in 1943–1945, these objects would have been so labeled."

Sightings

Foo fighters were reported on many occasions from around the world; a few examples are noted below.

Sighting from September 1941 in the Indian Ocean was similar to some later foo fighter reports. From the deck of the *S.S. Pułaski* (a Polish merchant vessel transporting British troops), two sailors reported a "strange globe glowing with greenish light, about half the size of the full moon as it appears to us." They alerted a British officer, who watched the object's movements with them for over an hour.

Charles R. Bastien of the Eighth Air Force reported one of the first encounters with foo fighters over the Belgium/Netherlands area; he described them as "two fog lights flying at high rates of speed that could change direction rapidly." During debriefing, his intelligence officer told him that two RAF night fighters had reported the same thing, and it was later reported in British newspapers.

Career U.S. Air Force pilot Duane Adams often related that he had witnessed two occurrences of a bright light which paced his aircraft for about half an hour and then rapidly ascended into the sky. Both incidents occurred at night, both over the South Pacific, and both were witnessed by the entire aircraft crew. The first sighting occurred shortly after the end of World War II while Adams piloted a B-25 bomber. The second sighting occurred in the early 1960s when Adams was piloting a KC-135 tanker.

Balls of Fire Stalk U. S. Fighters In Night Assaults Over Germany

By The Associated Press.

AMERICAN NIGHT FIGHTER BASE, France, Jan. 1—The Germans have thrown something new into the night skies over Germany —the weird, mysterious "foo-fighter," balls of fire that race alongside the wings of American Beaufighters flying intruder missions over the Reich.

American pilots have been encountering the eerie "foo-fighter" for more than a month in their night flights. No one apparently knows exactly what this sky weapon is.

The balls of fire appear suddenly and accompany the planes for miles. They appear to be radio-controlled from the ground and keep up with planes flying ...

Donald Meiers of Chicago said. "One is red balls of fire which appear off our wing tips and fly along with us; the second is a vertical row of three balls of fire which fly in front of us, and the third is a group of about fifteen lights which appear off in the distance—like a Christmas tree up in the air—and flicker on and off."

The pilots of this night-fighter squadron—in operation since September, 1943—find these fiery balls the weirdest thing that they have yet encountered. They are convinced that the "foo-fighter" is designed to be a psychological as well as a military weapon, although it is not the nature of the ...

A foo fighter photgraphed on the tail of this allied bomber. Why are they kept a secret to this day?

Explanations and Theories

Author Renato Vesco revived the wartime theory that the foo fighters were a Nazi secret weapon in his work *Intercept UFO*, reprinted in a revised English edition as *Man-Made UFOs: 50 Years of Suppression* in 1994. Vesco claims that the foo fighters were in fact a form of ground-launched automatically guided jet-propelled flak mine called the Feuerball (Fireball). The device, operated by special SS units, supposedly resembled a tortoise shell in shape, and flew by means of gas jets that spun like a Catherine wheel around

Foo fighters photgraphed around bombers in Italy, 1944.

the fuselage. Miniature klystron tubes inside the device, in combination with the gas jets, created the foo fighters' characteristic glowing spheroid appearance. A crude form of collision avoidance radar ensured the craft would not crash into another airborne object, and an onboard sensor mechanism would even instruct the machine to depart swiftly if it was fired upon. The purpose of the Feuerball, according to Vesco, was two-fold. The appearance of this weird device inside a bomber stream would (and indeed did) have a distracting and disruptive effect on the bomber pilots; and Vesco alleges that the devices were also intended to have an offensive capability. Electrostatic discharges from the klystron tubes would, he states, interfere with the ignition systems of the bombers' engines, causing the planes to crash. Although there is no hard evidence to support the reality of the Feuerball drone, this theory has been taken up by other aviation/ufology authors, and has even been cited as the most likely explanation for the phenomena in at least one recent television documentary on Nazi secret weapons. (End Wikipedia article)

So here we have the Wikipedia discussion on foo fighters, ending with the suggestion that foo fighters were possibly a Nazi secret weapon. But, if foo

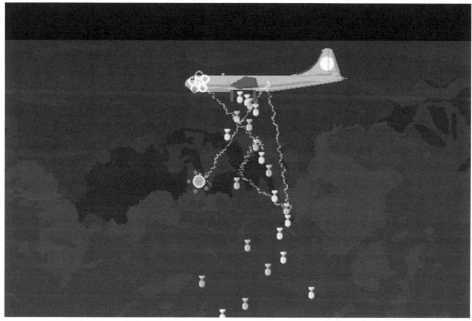

Foo fighters interfering with the engines and bombs of a bomber.

fighters were built by the Germans at the end of WWII, why would the US military, or the British for that matter, be keeping it a secret after so many years? Is it because the mercury gyro theory ultimately leads to the development and construction of not just foo fighters (Feuerballs) but also of larger mercury plasma craft such as the Vril and Haunebu flying saucer designs? These vessels apparently used a large mercury gyro in the center of the craft for the main lift and then three smaller mercury gyros around the base for steering and

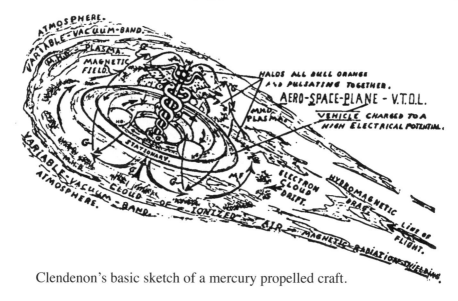

Clendenon's basic sketch of a mercury propelled craft.

direction of the craft.

Bill Clendenon, who first proposed that mercury gyros were used in foo fighters and flying saucers in the 1960s, thought that the fluttering of a hovering flying saucer just before taking off in one direction—something that he had witnessed himself—was the change in the stabilizing gyros as they were calibrated for the saucer to move in the desired direction.

Clendenon also maintained that these craft could also underwater and could be used as literal flying submarines. He and many researchers have noted the many cases in which a UFO either enters a body of water or emerges from it. Landing a flying saucer in a lake and parking it there for a day or two is a good way to hide a vehicle that one does not want to have spotted.

Ghost Rockets

A similar phenomenaon to foo fighters was the ghost rocket sightings immediately after WWII. Some ghost rockets were seen to land in lakes in Scandinavia and one lake was said to have an usual crater on the bottom that was discovered while searching for a ghost rocket seen to have landed in the lake. Let us look at what Wikipedia says about ghost rockets:

Ghost rockets (Swedish: Spökraketer, also called Scandinavian ghost rockets) were rocket- or missile-shaped unidentified flying objects sighted in 1946, mostly in Sweden and nearby countries. The first reports of ghost rockets were made on February 26, 1946, by Finnish observers. About 2,000 sightings were logged between May and December 1946, with peaks on 9 and 11 August 1946. Two hundred sightings were verified with radar returns, and authorities recovered physical fragments which were attributed to ghost rockets.

Investigations concluded that many ghost rocket sightings were probably caused by meteors. For example, the peaks of the sightings, on 9 and 11 August 1946, also fall within the peak of the annual Perseid meteor shower. However, most ghost rocket sightings did not occur during meteor shower activity, and furthermore displayed characteristics

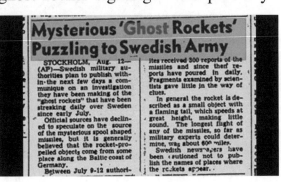

Mysterious 'Ghost Rockets' Puzzling to Swedish Army

STOCKHOLM, Aug. 12—(AP)—Swedish military authorities plan to publish within the next few days a communique on an investigation they have been making of the "ghost rockets" that have been streaking daily over Sweden since early July.

Official sources have declined to speculate on the source of the mysterious spool shaped missiles, but it is generally believed that the rocket-propelled objects come from some place along the Baltic coast of Germany.

Between July 9-12 authori-

ites received 300 reports of the missiles and since then reports have poured in daily. Fragments examined by scientists gave little in the way of clues.

In general the rocket is described as a small object with a flaming tail, which speeds at great height, making little sound. The longest flight of any of the missiles, so far as military experts could determine, was about 600 miles.

Swedish newspapers have been cautioned not to publish the names of places where the rockets appear.

inconsistent with meteors, such as reported maneuverability.

Debate continues as to the origins of the unidentified ghost rockets. In 1946, however, it was thought likely that they originated from the former German rocket facility at Peenemünde, and were long-range tests by the Russians of captured German V-1 or V-2 missiles, or perhaps another early form of cruise missile because of the ways they were sometimes seen to maneuver. This prompted the Swedish Army to issue a directive stating that newspapers were not to report the exact location of ghost rocket sightings, or any information regarding the direction or speed of the object. This information, they reasoned, was vital for evaluation purposes to the nation or nations performing the tests.

A ghost rocket photographed over Sweden in 1946.

Descriptions and Early Investigations

The early Russian origins theory was rejected by Swedish, British, and U.S. military investigators because no recognizable rocket fragments were ever found, and according to some sightings the objects usually left no exhaust trail, some moved too slowly and usually flew horizontally, they sometimes traveled and maneuvered in formation, and they were usually silent.

The sightings most often consisted of fast-flying rocket- or missile- shaped objects, with or without wings, visible for mere seconds. Instances of slower moving cigar shaped objects are also known. A hissing or rumbling sound was sometimes reported.

Crashes were not uncommon, almost always in lakes. Reports were made of objects crashing into a lake, then propelling themselves across the surface before sinking, as well as ordinary crashes. The Swedish military performed several dives in the affected lakes shortly after the crashes, but found nothing other than occasional craters in the lake bottom or torn off aquatic plants.

The best known of these crashes occurred on July 19, 1946, into Lake Kölmjärv, Sweden. Witnesses reported a gray, rocket-shaped object with wings crashing in the lake. One witness interviewed heard a thunderclap, possibly the object exploding. However, a 3-week military search conducted in intense secrecy again turned up nothing.

Immediately after the investigation, the Swedish Air Force officer who led the search, Karl-Gösta Bartoll submitted a report in which he stated that the bottom of the lake had been disturbed but nothing found and that "there are many indications that the Kölmjärv object disintegrated itself... the object was probably manufactured in a lightweight material, possibly a kind of magnesium alloy that would disintegrate easily, and not give indications on our instruments." When Bartoll was later interviewed in 1984 by Swedish researcher Clas Svahn, he again said their investigation suggested the object largely disintegrated in flight and insisted that "what people saw were real, physical objects."

On October 10, 1946, the Swedish Defense Staff publicly stated, "Most observations are vague and must be treated very skeptically. In some cases, however, clear, unambiguous observations have been made that cannot be explained as natural phenomena, Swedish aircraft, or imagination on the part of the observer. Echo, radar, and other equipment registered readings but gave no clue as to the nature of the objects." It was also stated that fragments alleged to have come from the missiles were nothing more than ordinary coke or slag.

On December 3, 1946, a memo was drafted for the Swedish Ghost Rocket committee stating, "nearly one hundred impacts have been reported and thirty pieces of debris have been received and examined by Swedish National Defense Research Institute (FOA)" (later said to be meteorite fragments). Of the nearly 1000 reports that had been received by the Swedish Defense Staff to November 29, 225 were considered observations of "real physical objects" and every one had been seen in broad daylight.

U.S. Involvement

In early August 1946 Swedish Lt. Lennart Neckman of the Defense Staff's Air Defense Division saw something that was "without a doubt ... a rocket projectile."

On August 14, 1946, the *New York Times* reported that Undersecretary of State Dean Acheson was "very much interested" in the ghost rocket reports, so was U.S. Army Air Force's intelligence as indicated nonpublicly by later documents (Clark, 246). Then on August 20, the *Times* reported that two U.S. experts on aerial warfare, aviation legend General Jimmy Doolittle and General David Sarnoff, president of RCA, arrived in Stockholm, ostensibly on private business and independently of each other. The official explanation was that Doolittle, who was now vice-president of the Shell Oil Company, was inspecting Shell branch offices in Europe, while Sarnoff, a former member of General Dwight D. Eisenhower's London staff, was studying the market for radio equipment. However, the *Times* story indicated that the Chief of the Swedish Defense Staff, made no secret that he "was extremely interested in asking the two generals advice and, if possible, would place all available reports before them." (Carpenter chronology) Doolittle and Sarnoff were briefed that on several occasions the ghost rockets had been tracked on radar. Sarnoff was later quoted by the *New York Times* on September 30 saying that he was "convinced that the 'ghost bombs' are no myth but real missiles."

On August 22, 1946, the director of the Central Intelligence Group (CIG), Lt. Gen. Hoyt Vandenberg, wrote a Top Secret memo to President Truman, perhaps based in part on information from Doolittle and Sarnoff. Vandenberg stated that the "weight of evidence pointed to Peenemünde as origin of the missiles, that US MA (military attaché) in Moscow had been told by 'key Swedish Air Officer' that radar course-plotting had led to conclusion that Peenemünde was the launch site. CIG speculates that the missiles are extended-range developments of V-1 being aimed for the Gulf of Bothnia for test purposes and do not overfly Swedish territory specifically for intimidation; self-destruct by small demolition charge or burning."

Nevertheless, there are no reports of rocket launches at Peenemünde or the Greifswalder Oie after February 21, 1945.

Although the official opinion of the Swedish and U.S. military remains unclear, a Top Secret USAFE (United States Air Force Europe) document from 4 November, 1948, indicates that at least some investigators believed the ghost rockets and later "flying saucers" had extraterrestrial origins. Declassified only in 1997, the document states:

For some time we have been concerned by the recurring reports on flying saucers. They periodically continue to pop up; during the last week, one was observed hovering over Neubiberg Air Base for about thirty minutes. They have been reported by so many sources and from such a variety of places that we are convinced that they cannot be disregarded and must be explained on some basis which is perhaps slightly beyond the scope of our present intelligence thinking.

When officers of this Directorate recently visited the Swedish Air Intelligence Service, this question was put to the Swedes. Their answer was that some reliable and fully technically qualified people have reached the conclusion that 'these phenomena are obviously the result of a high technical skill which cannot be credited to any presently known culture on earth.' They are therefore assuming that these objects originate from some previously unknown or unidentified technology, possibly outside the earth.

The document also mentioned a flying saucer crash search in a Swedish lake conducted by a Swedish naval salvage team, with the discovery of a previously unknown crater on the lake floor believed caused by the object (possibly referencing the Lake Kölmjärv search for a ghost rocket discussed above, though the date is unclear). The document ends with the statement that "we are inclined not to discredit entirely this somewhat spectacular theory [extraterrestrial origins], meantime keeping an open mind on the subject."

Greek Government Investigation

The "ghost rocket" reports were not confined to Scandinavian countries. Similar objects were soon reported early the following month by British Army units in Greece, especially around Thessaloniki. In an interview on September 5, 1946, the Greek Prime Minister, Konstantinos Tsaldaris, likewise reported a number of projectiles had been seen over Macedonia and Thessaloniki on September 1. In mid-September, they were also seen in Portugal, and then in Belgium and Italy. The Greek government conducted their own investigation, with their leading scientist, physicist Dr. Paul Santorinis, in charge. Santorinis had been a developer of the proximity fuze on the first A-bomb and held patents

2-5317.

TOP SECRET

| USAFE 14 | TT 1524 | TOP SECRET | 4 Nov 1948 |

From OI OB

For some time we have been concerned by the recurring reports on flying saucers. They periodically continue to cop up; during the last week, one was observed hovering over Neubiberg Air Base for about thirty minutes. They have been reported by so many sources and from such a variety of places that we are convinced that they cannot be disregarded and must be explained on some basis which is perhaps slightly beyond the scope of our present intelligence thinking.

When officers of this Directorate recently visited the Swedish Air Intelligence Service. This question was put to the Swedes. Their answer was that some reliable and fully technically qualified people have reached the conclusion that "these phenomena are obviously the result of a high technical skill which cannot be credited to any presently known culture on earth." They are therefore assuming that these objects originate from some previously unknown or unidentified technology, possibly outside the earth.

One of these objects was observed by a Swedish technical expert near his home on the edge of a lake. The object crashed or landed in the lake and he carefully noted its azimuth from his point of observation. Swedish intelligence was sufficiently confident in his observation that a naval salvage team was sent to the lake. Operations were underway during the visit of USAFE officers. Divers had discovered a previously uncharted crater on the floor of the lake. No further information is available, but we have been promised knowledge of the results. In their opinion, the observation was reliable, and they believe that the depression on the floor of the lake, which did not appear on current Hydrographic charts, was in fact caused by a flying saucer.

Although accepting this theory of the origin of these objects poses a whole new group of questions and puts much of our thinking in a changed light, we are inclined not to discredit entirely this somewhat spectacular theory, meantime keeping an open mind on the subject. What are your reactions?

TOP SECRET

(END OF USAFE ITEM 14)

on guidance systems for Nike missiles and radar systems. Santorinis was supplied by the Greek Army with a team of engineers to investigate what again were believed to be Russian missiles flying over Greece.

In a 1967 lecture to the Greek Astronomical Society, broadcast on Athens Radio, Santorinis first publicly revealed what had been found in his 1947 investigation. "We soon established that they were not missiles. But, before we could do any more, the Army, after conferring with foreign officials (presumably U.S. Defense Dept.), ordered the investigation stopped. Foreign scientists [from Washington] flew to Greece for secret talks with me." Later Santorinis told UFO researchers such as Raymond Fowler that secrecy was invoked because officials were afraid to admit of a superior technology against which we have "no possibility of defense." (End Wikipedia Article)

The Quantum Vacuum Thruster

When discussing mercury gyros and plasmas one final device should be mentioned: the quantum vacuum thruster. We will end this chapter with the Wikipedia article on the quantum vacuum thruster:

A quantum vacuum thruster (QVT or Q-thruster) is a theoretical system that uses the same principles and equations of motion that a conventional plasma thruster would use, namely magnetohydrodynamics (MHD), to make predictions about the behavior of the propellant. However, rather than using a conventional plasma as a propellant, a QVT uses the quantum vacuum fluctuations of the zero-point field as the fuel source. If QVT systems were to truly work they would eliminate the need to carry any propellant, as the system uses the quantum vacuum to assist with thrust. It would also allow for much higher specific impulses for QVT systems compared to other spacecraft as they would be limited only by their power supply's energy storage densities. Harold White's Advanced Propulsion Physics Laboratory (NASA Eagleworks) suggests that their RF cavity may be an example of a quantum vacuum thruster (QVT or Q-thruster).

History and Controversy

The name and concept is controversial. In 2008, Yu Zhu and others at China's

Northwestern Polytechnical University claimed to measure thrust from such a thruster, but called it a "microwave thruster without propellant" working on quantum principles. In 2011 it was mentioned as something to be studied by Harold G. White and his team at NASA's Eagleworks Laboratories, who were working with a prototype of such a thruster. Other physicists, such as Sean M. Carroll and John Baez, dismissed it because the quantum vacuum as currently understood is not a plasma and does not possess plasma-like characteristics.

Theory of Operation

A vacuum can be viewed not as empty space but as the combination of all zero-point fields. According to quantum field theory the universe is made up of matter fields whose quanta are fermions (e.g. electrons and quarks) and force fields, whose quanta are bosons (i.e. photons and gluons). All these fields have some intrinsic zero-point energy. Describing the quantum vacuum, a *Physics Today* article cited by the NASA team describes this ensemble of fields as "a turbulent sea, roiling with waves associated with a panoply of force-mediating fields such as the photon and Higgs fields." Given the equivalence of mass and energy expressed by Einstein's $E = mc2$, any point in space that contains energy can be thought of as having mass to create particles. Virtual particles spontaneously flash into existence and annihilate each other at every point in space due to the energy of quantum fluctuations. Many real physical effects attributed to these vacuum fluctuations have been experimentally verified, such as spontaneous emission, Casimir force, Lamb shift, magnetic moment of the electron and Delbrück scattering; these effects are usually called "radiative corrections."

The Casimir effect is a weak force between two uncharged conductive plates caused by the zero-point energy of the vacuum. It was first observed

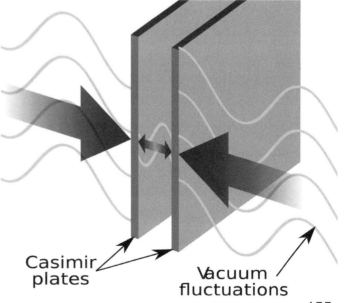

Casimir plates

Vacuum fluctuations

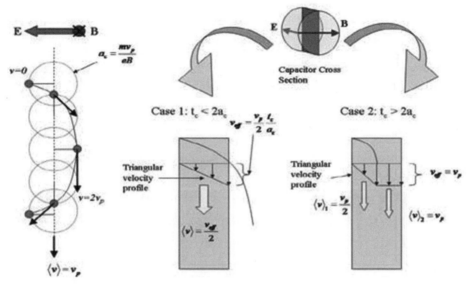

The Q-Thruster theory.

experimentally by Lamoreaux (1997) and results showing the force have been repeatedly replicated. Several scientists including White have highlighted that a net thrust can indeed be induced on a spacecraft via the related "dynamical Casimir effect." The dynamic Casimir effect was observed experimentally for the first time in 2011 by Wilson et al. In the dynamical Casimir effect electromagnetic radiation is emitted when a mirror is accelerated through space at relativistic speeds. When the speed of the mirror begins to match the speed of the photons, some photons become separated from their virtual pair and so do not get annihilated. Virtual photons become real and the mirror begins to produce light. This is an example of Unruh radiation. A publication by Feigel (2004) raised the possibility of a Casimir-like effect that transfers momentum from zero-point quantum fluctuations to matter, controlled by applied electric and magnetic fields. These results were debated in a number of follow up papers in particular van Tiggelen et al (2006) found no momentum transfer for homogeneous fields, but predict a very small transfer for a Casimir-like field geometry. This cumulated with Birkeland & Brevik (2007) who showed that electromagnetic vacuum fields can cause broken symmetries (anisotropy) in the transfer of momentum or, put another way, that the extraction of momentum from electromagnetic zero-point fluctuations is possible in an analogous way that the extraction of energy is possible from the Casimir effect.

156

Birkeland & Brevik highlight that momentum asymmetries exist throughout nature and that the artificial stimulation of these by electric and magnetic fields have already been experimentally observed in complex liquids. This relates to the Abraham–Mikowski controversy, a long theoretical and experimental debate that continues to the current time. It is widely recognized that this controversy is an argument about definition of the interaction between matter and fields. It has been argued that momentum transfer between matter and electromagnetic fields relating to the Abraham-Minikowski issue would allow for propellantless drives.

A QVT system seeks to make use of this predicted Casimir-like momentum transfer. It is argued that when the vacuum is exposed to crossed electric and magnetic fields (i.e. E and B-fields) it will induce a drift of the entire vacuum plasma which is orthogonal to that of the applied E x B fields. In a 2015 paper White highlighted that the presence of ordinary matter is predicted to cause an energy perturbation in the surrounding quantum vacuum such that the local vacuum state has a different energy density when compared with the "empty" cosmological vacuum energy state. This suggests the possibility of modeling the vacuum as a dynamic entity as opposed to it being an immutable and non-degradable state. White models the perturbed quantum vacuum around a hydrogen atom as a Dirac vacuum consisting of virtual electron-positron pairs. Given the nontrivial variability in local energy densities resulting from virtual pair production he suggests the tools of magnetohydrodynamics (MHD) can be used to model the quasiclassical behavior of the quantum vacuum as a plasma.

White compares changes in vacuum energy density induced by matter to the hypothetical chameleon field or quintessence currently being discussed in the scientific literature. It is claimed the existence of a "chameleon" field whose mass is dependent on the local matter density may be an explanation for dark energy. A number of notable physicists, such as Sean Carroll, see the idea of a dynamical vacuum energy as the simplest and best explanation for dark energy. Evidence for quintessence would come from violations of Einstein's equivalence principle and variation of the fundamental constants ideas which are due to be tested by the Euclid telescope which is set to launch in 2020.

Systems utilizing Casimir effects have thus far been shown to only create very small forces and are generally considered one-shot devices that would require a subsequent energy to recharge them (i.e. Forward's "vacuum fluctuation battery"). The ability of systems to use the zero-point field continuously as a source of energy or propellant is much more contentious (though peer-reviewed models have been proposed). There is debate over which formalisms of quantum mechanics apply to propulsion physics under such circumstances, the more refined Quantum Electrodynamics (QED), or the relatively undeveloped and controversial Stochastical Quantum Electrodynamics (SED). SED describes electromagnetic energy at absolute zero as a stochastic, fluctuating zero-point field. In SED the motion of a particle immersed in the stochastic zero-point radiation field generally results in highly nonlinear behavior. Quantum effects emerge as a result of permanent matter-field interactions not possible to describe in QED. The typical mathematical models used in classical electromagnetism, quantum electrodynamics (QED) and the standard model view electromagnetism as a U(1) gauge theory, which topologically restricts any complex nonlinear interaction. The electromagnetic vacuum in these theories is generally viewed as a linear system with no overall observable consequence. For many practical calculations zero-point energy is dismissed by fiat in the mathematical model as a constant that may be canceled or as a term that has no physical effect.

The 2016 NASA paper highlights that stochastic electrodynamics (SED) allows for a pilot-wave interpretation of quantum mechanics. Pilot-wave interpretations of quantum mechanics are a family of deterministic nonlocal theories distinct from other more mainstream interpretations such as the Copenhagen interpretation and Everett's many-worlds interpretation. Pioneering experiments by Couder and Fort beginning in 2006 have shown that macroscopic classical pilot-waves can exhibit characteristics previously thought to be restricted to the quantum realm. Hydrodynamic pilot-wave analogs have been able to duplicate the double slit experiment, tunneling, quantized orbits, and numerous other quantum phenomena and as such pilot-wave theories are experiencing a resurgence in interest. Coulder and Fort note in their 2006 paper that pilot-waves are nonlinear dissipative systems sustained by external forces. A dissipative system is characterized by the spontaneous appearance of symmetry breaking (anisotropy) and the formation of complex, sometimes chaotic or emergent, dynamics where interacting fields can exhibit long range correlations. In SED the zero point field (ZPF) plays the role of the pilot wave that guides real particles on their way. Modern approaches to SED consider wave and particle-like quantum effects as well-coordinated emergent systems that are the result of speculated sub-quantum interactions with the zero-point field.

Controversy and Criticism

A number of notable physicists have found the Q-thruster concept to be implausible. For example, mathematical physicist John Baez has criticized the reference to "quantum vacuum virtual plasma" noting that: "There's no such thing as 'virtual plasma.'" Noted Caltech theoretical physicist Sean M. Carroll has also affirmed this statement, writing "[t]here is no such thing as a 'quantum vacuum virtual plasma,'..." In addition, Lafleur found that quantum field theory predicts no net force, implying that the measured thrusts are unlikely to be due to quantum effects. However, Lafleur noted that this conclusion was based on the assumption that the electric and magnetic fields were homogeneous, whereas certain theories posit a small net force in inhomogeneous vacuums.

Especially, the violation of energy and momentum conservation laws have been heavily criticized. In a presentation at NASA Ames Research Center in November, 2014, Harold White addressed the issue of conservation of momentum by stating that the Q-thruster conserves momentum by creating a wake or anisotropic state in the quantum vacuum. White indicated that once false positives were ruled out, Eagleworks would explore the momentum distribution and divergence angle of the quantum vacuum wake using a second Q-thruster to measure the quantum vacuum wake. In a paper published in January, 2014, White proposed to address the conservation of momentum issue by stating that the Q-thruster pushes quantum particles (electrons/positrons) in one direction, whereas the Q-thruster recoils to conserve momentum in the other direction. White stated that this principle was similar to how a submarine uses its propeller to push water in one direction, while the submarine recoils to conserve momentum. Hence, the violations of fundamental laws of physics can be avoided.

Other Hypothesized Quantum Vacuum Thrusters

A number of physicists have suggested that a spacecraft or object may generate thrust through its interaction with the quantum vacuum. For example,

Fabrizio Pinto in a 2006 paper published in the *Journal of the British Interplanetary Society* noted it may be possible to bring a cluster of polarisable vacuum particles to a hover in the laboratory and then to transfer thrust to a macroscopic accelerating vehicle. Similarly, Jordan Maclay in a 2004 paper titled "A Gedanken Spacecraft that Operates Using the Quantum Vacuum (Dynamic Casimir Effect)" published in the scientific journal *Foundations of Physics* noted that it is possible to accelerate a spacecraft based on the dynamic Casimir effect, in which electromagnetic radiation is emitted when an uncharged mirror is properly accelerated in vacuum. Similarly, Puthoff noted in a 2010 paper titled "Engineering the Zero-Point Field and Polarizable Vacuum For Interstellar Flight" published in the *Journal of the British Interplanetary Society* noted that it may be possible that the quantum vacuum might be manipulated so as to provide energy/thrust for future space vehicles. Likewise, researcher Yoshinari Minami in a 2008 paper titled "Preliminary Theoretical Considerations for Getting Thrust via Squeezed Vacuum" published in the Journal of the British Interplanetary Society noted the theoretical possibility of extracting thrust from the excited vacuum induced by controlling squeezed light. In addition, Alexander Feigel in a 2009 paper noted that propulsion in quantum vacuum may be achieved by rotating or aggregating magneto-electric nano-particles in strong perpendicular electrical and magnetic fields.

However, according to Puthoff, although this method can produce angular momentum causing a static disk (known as a Feynman disk) to begin to rotate, it cannot induce linear momentum due to a phenomenon known as "hidden momentum" that cancels the ability of the proposed E×B propulsion method to generate linear momentum. However, some recent experimental and theoretical work by van Tiggelen and colleagues suggests that linear momentum may be transferred from the quantum vacuum in the presence of an external magnetic field. (End Wikipedia Article)

Chapter 9

Electrogravitics For Advanced Propulsion
By Thomas Valone, M.A., P.E.

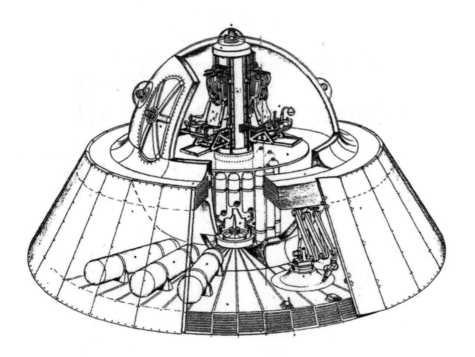

ELECTROGRAVITICS FOR ADVANCED PROPULSION

By Thomas Valone, M.A., P.E.
Integrity Research Institute

Recently, two 1956 military documents, "Electrogravitics Systems" and "The Gravitics Situation," originally published by the Gravity Research Group of London (Special Weapons Study Unit), were declassified. Outlining T. Townsend Brown's antigravity discovery (see *Atlantis Rising*, Number 22, p.35; *AIR International*, Jan., 2000; *Jane's Defence Weekly*, 10 June 1995, p.34), and the subsequent Project Winterhaven, they were a vital new chapter in aviation research. For example, the documents state, "Unlike the turbine engine, electrogravitics is not just a new propulsion system, it is a new mode of thought in aviation and communications, and it is something that may become all-embracing."

To explain, "electrogravitics" is the science of using high voltage electricity to provide propulsive force to aircraft or spacecraft of certain geometries. Or as Jeane Manning explains, "The apparatus is pulled along by its self-generated gravity field, like a surfer riding a wave." Its discovery is often credited to Thomas Townsend Brown, a physicist who was encouraged by his professor, Dr. Paul Biefield, a former classmate of Albert Einstein. However, there are those who say that Professor Francis Nipher's experiments, electrically charging lead balls, published in the *Electrical Experimenter*, in 1918, predates Biefield/Brown. Unknown to many unconventional propulsion experts, T. Townsend Brown's electrogravitics work after the war involved a multinational project. American companies such as Douglas, Glenn Martin, General Electric, Bell, Convair, Lear, and Sperry-Rand participated in the research effort. Countries such as Britain, France, Sweden, Canada, and Germany also had concurrent projects from 1954 through 1956.

Furthermore, through the investigative effort of Dr. Paul LaViolette, it has become clear that electrogravitics became an integral part of the B-2 Stealth Bomber today, giving it an unlimited range. LaViolette challenges us with the question, "Could the B-2 really be the realization of one of mankind's greatest dreams—an aircraft that has mastered the ability to control gravity?" LaViolette's investigation is summarized in an article "The U. S. Antigravity Squadron" which has been reprinted, along with both reports mentioned above, in the book, *Electrogravitics Systems, A New Propulsion Methodology*. LaViolette's book, *Subquantum Kinetics: The Alchemy of Creation* includes a chapter on the theory of electrogravitics and a plot of applied voltage versus disc speed from Naval Research Lab data, which starts around 40 kilovolts and 2 miles per hour.

163

T. TOWNSEND BROWN

A curious fact revealed in T.T. Brown's first article "How I Control Gravity" (*Science and Invention*, 1929) is the alignment of the "molecular gravitors." These massive dielectrics provided the most propulsive force when the "differently charged elements" were aligned (with the voltage source). This sounds like crystal plane alignment and perhaps explains the article "Gravity Nullified: Quartz Crystals Charged by High Frequency Currents Lose Their Weight" which appeared two years earlier in the same magazine in September of 1927. The editors had a change of heart however, in the following issue, and rescinded the article.

T.T. Brown's first patent, #1,974,483 issued in 1934, "Electrostatic Motor," is a fascinating free energy machine as well as a propulsion source. Claiming an efficiency of a "million to one", Brown causes the massive dielectrics to be the workhorse of the motor, exceeding, in his words, "the well known pin wheel effect or reaction from a high voltage

T. Townsend Brown in his laboratory flying discs in his experiments.

point discharge." Much of what we know about T.T. Brown is from his numerous patents (all of them are reprinted in the *Electrogravitics Systems* book), although I was fortunate enough to correspond with him in 1981 when he was at the University of Florida. A sample of his detailed correspondence is contained in the book, *Ether-Technology: A Rational Approach to Gravity-Control* by Rho Sigma (1977) which is the only other introduction to Brown's work. The important fact from that book is that the DC power supply went up to 250 kV, with a substantial force being displayed starting around 150 kV. Here we get an idea of the range of voltage necessary for successful electrogravitics that even recent military contractors mysteriously disregard. An example is R. L. Talley's report to the Air Force concerned "with exploring the Biefield-Brown effect which allegedly converts electrostatic energy directly into a propulsive force in a vacuum environment." It was

164

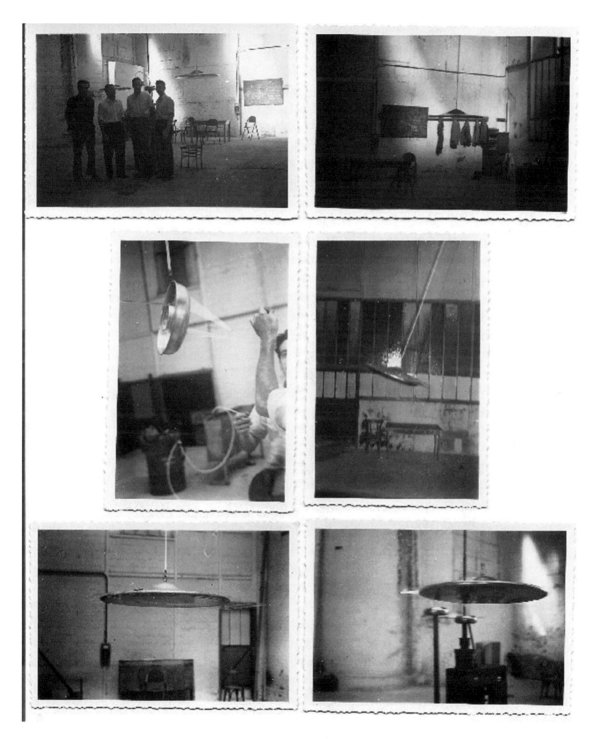

T. Townsend Brown in his laboratory flying discs in his experiments.

entitled, "Twenty First Century Propulsion Concept" #PL-TR-91-3009, but only tested Brown saucer designs in the range of 19 kV and predictably failed to produce results.

Brown's saucer tests show a propulsive force with the positive voltage leading and the negative edge trailing. The high voltage electrically charged the air around the craft with a cloud of positive ions forming in front of the craft and a cloud of negative ions behind. This has been verified with tests recently performed by researcher Larry Davenport. These tests are reprinted in the book, *Electrogravitics Systems* and can be seen in the commercial video, "Free Energy, The Race to Zero Point" for which I was the technical consultant.

In March, 1952, the Townsend Brown Foundation laboratory in Hollywood, California was visited by Air Force Major General Victor E. Bertrandias. He reported to Lt. Gen. H. A. Craig that he was "frightened" by the flying demonstration because it was in private hands and he felt it was "in the stage in which the atomic development was in the early days." He expressed concern about "if it ever gets away," meaning, we presume, "into enemy hands." A confidential security investigation was thereby initiated concerning the Foundation and T.T. Brown. Shortly afterwards, an evaluation by the Office of Naval Research in September, 1952 surprisingly devalued the Brown saucers to a "well-known phenomenon of the electric wind" claiming it would perform "less at high voltage and zero in a vacuum." The Navy declassified the report in October, 1952. However, today, copies of the report are not available from the Naval Research Laboratory in Washington, DC.

In 1956, Brown returned from a research trip in France where he verified that electrogravitics worked well in a vacuum, in other words, the environment of outer space. *Interavia Magazine* published an article in the same year about Brown entitled, "Towards flight without stress or strain or weight" and alluded to possible speeds of several hundred miles per hour. In 1958, *Fate* magazine writer Gaston Burridge described Brown's metal discs that reached up to 30 inches in diameter. Until 1960, Brown and Agnew Bahnson worked on various designs in Bahnson's laboratory which were recorded by Bahnson's daughter on Super-8 film. Today, a VHS converted silent video is available of those experiments entitled, "Thomas Townsend Brown: Bahnson Lab 1958-1960." In 1964 Bahnson, an experienced pilot, mysteriously flew into electric wires and died. The Bahnson heirs subsequently dissolved the laboratory project.

ELECTROGRAVITICS PUZZLE

In 1985, Dr. Paul LaViolette was in the Library of Congress in Washington, DC and looked up the work "gravity" in the card catalog. Surprisingly, he found the listing for "Electrogravitics Systems," a report that was missing from the stacks. When the librarian tried to locate any other copies through interlibrary loan, she commented, "It must be an exotic document" because she could find only one in the country which was at Wright-Patterson Air Force Base. Thus, LaViolette was successful in obtaining a copy of the formerly classified document. The mystery continued: seven years later when contacting the Wright-Patterson AFB Technical Library, they surprisingly found no reference in the computer-based card catalog. They did locate the document on the shelves, however, after being asked to search for it. To summarize, the report has historic value because:

- ♦ It validates T.T. Brown's experiments;
- ♦ It lists the major corporations that were collaborating on electrogravitics;
- ♦ It includes the requirements for supersonic speed;
- ♦ It shows the continuity from Project Winterhaven in 1952;
- ♦ The report includes a list of electrostatic patents;
- ♦ It had been classified by the Air Force for an undetermined amount of time which underscores its importance.

Prepared by the Aviation Studies (International) Ltd., Gravity Research Group, Special Weapons Study Unit in England in February of 1956, it defines electrogravitics as "a synthesis of electrostatic energy use for propulsion." The report historically notes that: "Electrogravitics had its birth after the War, when Townsend Brown sought to improve on the various proposals that then existed for electrostatic motors sufficiently to produce some visible manifestation of sustained motion." As mentioned in the first section of the report, both Project Winterhaven (1952) and "Electrogravitics Systems" (1956) propose "a saucer as the basis of a possible interceptor with Mach 3 capability." Another interesting detail presented is the necessity of an insulator with a exceedingly high "dielectric" constant of 30,000 for supersonic speed when the best dielectrics of that era were around 5,000. This section goes on to describe the creation of a <u>local gravitational system</u> by the craft which "would confer upon the fighter the sharp-edged changes of direction typical of motion in space." The January, 1955 entry states:

Back in 1948 and 49, the public in the U.S. had a surprisingly clear idea what a flying saucer should, or could do. There has never at any time been any realistic explanation of what propulsion agency could make it do those things, but its ability to move within its own gravitation field was presupposed from its manoeuverability. Yet all this was at least two years before electro-static energy was shown to produce propulsion. It is curious that the public were so far ahead of the empiricists on this occasion…

The intriguing part of this commentary is that without any space program at the time, the report complains that the public knows how UFOs behave and refers to sharp-edged changes of direction.

Later in the report, we read, "One of the difficulties in 1954 and 1955 was to get aviation to take electrogravitics seriously." However, corporations such as Douglas, Sperry, Bell, GE, Hiller, Lear, and Convair are then described with an ongoing-project perspective. For example we read that, "General Electric is working on the use of electronic rigs to make adjustments to gravity." "Glenn Martin say gravity control could be achieved in six years… Clarke Electronics state they have a rig, and add that in their view the source of gravity's force will be understood sooner than some people think." This information makes the report exciting reading and gives it an air of suspense.

Even today, electrogravitics continues to attract public attention in the press. The latest is an article entitled "Military Power" published in a British aviation magazine, *AIR International*, (Jan., 2000) that includes copies of LaViolette's drawings from the *Electrogravitics Systems* book. The article also cites the *Aviation Week and Space Technology* article from March, 1992 "Black world engineers, scientist, encourage using highly classified technology for civil applications" which caused LaViolette to investigate the B-2 Bomber connection to T.T. Brown's electrogravitics.

JOHN SEARL'S ELECTROGRAVITY

John R.R. Searl, of England, constructed numerous craft purported to fly with high voltage (see the recent biography *Antigravity: The Dream Made Reality* book by John Thomas). However, one correction to the "Antigravity" article from *Atlantis Rising* Number 22 issue is that the <u>positive</u> pole was traditionally at the periphery of Searl's crafts. This is important because as Searl describes his control of the imbalance of positive voltage on the edges, to steer the craft, he found that the saucers would travel toward the <u>more positive side</u>, exactly like T.T. Brown's saucers behave! Throughout the sixties and the seventies, J.R.R. Searl produced many newsletters detailing the work he was doing. Since I corresponded with him in 1981, I also received some of these reports. The importance of his experiments lies in the electrogravitics phenomena associated with them. In the 6/1/68 issue of the "Searl National Space Research Consortium" newsletter, Barrett reports that the ionization of the air and permanent electric polarity of dielectrics were common along with the antigravity effects. In the 6/14/71 issue of the newsletter, Bernhard Vaegs reports that "a pink halo surrounded the craft" and describes the effect of the millions of volts that were generated. This type of description is found throughout the reports and probably was measured by the length of the spark discharge considering the approximate voltage breakdown of air. Barrett describes in the 6/1/68 issue a vacuum layer that surrounds the craft preventing ionizing breakdown of the air. The similarities between Searl's high voltage propulsion and T.T. Brown's high voltage propulsion that both are based upon the principles of electrogravitics as theoretically predicted by Dr. Paul LaViolette in his previously mentioned book, *SubQuantum Kinetics*.

THE HUTCHISON EFFECT

In 1980, George Hathaway, a professional engineer licensed in Canada, along with entrepreneur, Alex Pizzaro, formed a small company to develop and promote what is referred to as "The Hutchison Effect." It is named after its inventor, John Hutchison, who liked to experiment with combinations of Tesla coils and Van de Graaff generators at the same time. Much of the information about the "lift and disruption" effects has been reported at various

168

conferences (such as in the *Third International Symposium on Non-Conventional Energy Technology* held in Hull, Quebec in 1986). Videotapes of much of the phenomena have been shown on Japanese TV as well. Hathaway also assembled a three-hour videotape that documents the TV interviews, reports, and actual events. To summarize, the experiments were conducted with 250 KV of DC power on the Van de Graaff and about the same voltage of AC power on the Tesla coil. The total real power was about 1.5 KW continuously, according to Hathaway. Besides the disruptive effects, which were numerous, the lifting of various heavy objects by the field was most impressive. These events can be seen at the end of the commercial video, "Free Energy, The Race to Zero Point." In regards to the AC contribution to the field, Hathaway reports that he measured a small voltage of 2 millivolts per meter in the active region (besides the DC offset). This is a small AC signal but on top of the high voltage DC signal, it performs amazing feats.

The importance of the Hutchison Effect to Brown electrogravitics is the AC "ripple" on the high DC voltage. A reference to this may be found in a military report by Dr. Dennis Cravens who gave T.T. Brown a high rating of "practicality." Cravens reported in his evaluation of Brown that older, high voltage supplies always had some AC ripple to the regulated signal, and wonders if this had any effect on Brown's phenomena (Cravens, T.L. "Electric Propulsion Study", AL-TR-89-040, #ADA 227121, Science Applic. Inter. Corp., Torrance, CA 90501). Dr. LaViolette has also found this factor to have particular electrogravitic significance.

THE B-2 STEALTH BOMBER CONNECTION

Thanks to Dr. Paul LaViolette reporting in his article, "The U.S. Antigravity Squadron", there is substantial evidence that the electrogravitics research of the 1950's actually resulted in the B-2 Stealth Bomber "auxiliary propulsion system." Summarizing Dr. LaViolette's article, with references cited therein, the following facts are the most convincing:

- ➤ The B-2 charges the leading edges of its wing-like body, with high voltage;
- ➤ The B-2 is shaped just like T.T. Brown suggested an electrogravitic craft should look, for maximum charge separation;
- ➤ Northrup tested leading-edge charging in 1968;
- ➤ B-2 electrically charges the exhaust similar to the suggestion by T.T. Brown that the craft should be powered by needle-point flame-jet generators which electrically charge the exhaust;
- ➤ *Aviation Week* admits the existence of "dramatic, classified technologies" applicable to "aircraft control and propulsion" on the B-2;
- ➤ *Aviation Week* also disclosed that a high-density dielectric ceramic RAM made of powdered depleted uranium happens to have a mass density of 3 times that of the high-K dielectrics tested in the 1950's;
- ➤ The B-2's Emergency Power Units (EPU) can drive an electrical generator at

   high altitudes or even in space, since the fuel can be made to decompose rapidly even without oxygen;

➤ Edward Aldridge, the Secretary of the Air Force, admits that the B-2 creates no vapor trail at high altitudes.

➤ The decomposed gases from the EPU's can function as the ion-carrying medium, in much the same way as the hot exhaust gases from the air breathing flame jet generators.

These details create a sense of excitement about the world's foremost aircraft. Dr. LaViolette argues that the electrogravitic drive will function better at higher speeds due to the better flow of the ions. Therefore, it is likely, he says, that the B-2 actually is a supersonic aircraft, especially since the 1968 leading-edge charging experiments were for supersonic softening of the shock wave.

House Representative Robert Walker was quoted recently in *Popular Science* ("Secrets of Groom Lake") as promoting the idea of declassifying military secrets that will help commercial development. We hope that this trend will continue so that advanced Shuttle designs may also acquire an electrogravitic drive.

Through a proposal he submitted in 1990, Dr. LaViolette made NASA aware that an electrogravitic drive would be a feasible propulsion method for the Mars journey because calculations show that the transit time can be less than one month, instead of half a year to a year. It is especially attractive since it uses so little power when it is operational and verified by Brown to work well in a vacuum. More information about electrogravitics and Brown's patents are available in my book, *Electrogravitics Systems: Reports on a New Propulsion Methodology*. Many of the books and videos mentioned in this article are available, as a public education service, from our non-profit organization, Integrity Research Institute, www.integrity-research.org

Empirical Analysis of Electrogravitics and Electrokinetics and its Potential for Space Travel

Thomas F. Valone[*]
Integrity Research Institute, Beltsville MD 20705
www.IntegrityResearchInstitute.org
e-mail: *IRI@starpower.net*

An analysis of the 90-year old science of electrogravitics (a.k.a. "gravitics" or "electrogravity") necessarily includes an analysis of electrokinetics. Electrogravitics is most commonly associated with the 1928 British patent #300,311 of T. Townsend Brown (his first one), the 1952 Special Inquiry File #24-185 of the Office of Naval Research into the "Electro-Gravity Device of Townsend Brown" and two widely circulated 1956 Aviation Studies Ltd. reports on "Electrogravitics Systems" and "The Gravitics Situation." By definition, electrogravitics historically has had a purported relationship to gravity or the object's mass, as well as the applied voltage. The Gravitics Situation report defined electrogravitics as "The application of modulating influences on electrostatic propulsion system." It also was tested recently by the Honda Corporation, which published experimental results and proposed theory of a correlation between electricity and gravity. Electrokinetics, on the other hand, is more commonly associated with many later patents of T. Townsend Brown as well as Agnew Bahnson, starting with the 1960 US patent #2,949,550 entitled, "Electrokinetic Apparatus." Electrokinetics, which often involves a capacitor and dielectric, has virtually no relationship that can be connected with mass or gravity. The Army Research Lab has recently issued a report on electrokinetics, analyzing the force on an asymmetric capacitor, while NASA has received three patents on the same design topic. To successfully describe and predict the reported motion toward the positive terminal of the capacitor, it is desirable to use the classical electrokinetic field and force equations for the specific geometry involved. This initial review and analysis also suggests directions for further confirming experiments and an empirically-based formulation of a working hypothesis for electrokinetics.

I. Nomenclature

J	=	electric current density
I	=	electric current
E_K	=	electrokinetic force vector
B	=	magnetic flux density
E	=	electric field
ρ	=	charge density

II. Introduction to Electrogravitics versus Electrokinetics

FOURTEEN years ago the first edited volume on the subject, *Electrogravitics Systems Volume I: A New Propulsion Methodology* or just "Volume I", introduced the subject by reprinting the Aviation Studies reports from 1956 as well as an in-depth analysis of the B-2 bomber by Paul LaViolette.[1] The second volume, *Electrogravitics II: Validating Reports on a New Propulsion Methodology* or "Volume II" expands the historical perspective of the first volume and brings it up to date. For example, Volume II contains further information on the Army Research Lab and Honda Corporation experiments, as well as the electrokinetic equation discovery presented in this paper. A short review of the history of electrogravitics has recently been published by Professor Theodore Loder.[2]

---

[*] President, Integrity Research Institute, 5020 Sunnyside Avenue, Suite 209, Beltsville MD 20705, AIAA Member.

The Anti-Gravity Files

A working definition, based on the T. Townsend Brown's first patent #300,311 and The Gravitics Situation report is "*electricity used to create a force that depends upon an object's mass, similar to gravity.*" This is the answer that perhaps should still be used to identify true electrogravitics, which also involves the object's mass in the force, often with a dielectric. This is also what the "Biefeld-Brown effect" of describes. However, we have seen T. Townsend Brown and his patents evolve over time which Tom Bahder emphasizes. Later on, Brown refers to "electrokinetics" (that partly overlaps the field of electrogravitics), that requires asymmetric capacitors to amplify the force. Therefore, Bahder's article discusses the lightweight effects of "lifters" and the ion mobility theory found to explain them. Note: *electrogravitics (EG) and electrokinetics (EK) are related but different phenomena.*

To put things in perspective, the article "How I Control Gravitation," published in 1929 by Brown,[3] presents an electrogravitics-validating discovery about *very heavy metal objects* (44 lbs. each) separated by an insulator, charged up to high voltages. T.T. Brown also expresses an experimental formula in words which tell us what he found was directly contributing to the *unidirectional force* (UDF) which he discovered, moving the system of masses toward the positive charge. He describes the equation for his electrogravitic force to be $F \approx V m_1 m_2 / r^2$. However, electrokinetics and electrogravitics also seem to be governed by another equation (Eq.1) when higher order pulsed voltages are utilized .

A. Zinsser Effect versus the Biefeld-Brown Effect

To expand and support the empirical evidence for electrokinetics, there is another invention which has comparable experiments that also involve electrogravity, called "gravitational anisotropy" by Rudolf G. Zinsser from Germany. Zinsser presented his experimental results at the Gravity Field Conference in Hanover in 1980, and also at the First International Symposium of Non-Conventional Energy Technology in Toronto in 1981.[4] For years afterwards, all of the scientists who knew of Zinsser's work, including myself, regarded his invention as a unique phenomenon, not able to be classified with any other discovery. However, upon comparing Zinsser to Brown's 1929 article on gravitation referred to above, there are striking similarities.

Zinsser's discovery is detailed in *The Zinsser Effect* book by this author.[5] To summarize his life's work, Zinsser discovered that if he connected his patented pulse generator to two conductive metal plates immersed in water, he could induce a sustained force that lasted even after the pulse generator was turned off. The pulses lasted for only a few nanoseconds each.[6] Zinsser called this input "a kinetobaric driving impulse." Furthermore, he points out in the Specifications and Enumerations section, that the high dielectric constant of water (about 80) is desirable and that a solid dielectric is possible. Dr. Peschka calculated that Zinsser's invention produced 6 Ns/Ws or 6 N/W.[7] This figure is *twenty times* the force per energy input of the Inertial Impulse Engine of Roy Thornson, (report available from IRI) which has been estimated to produce 0.32 N/W.[8] By comparison, it is important to realize that any production of force today is less efficient, as seen by the fact that a DC-9 jet engine produces *about 20 times less:* only 0.016 N/W or 3 lb/hp (fossil-fuel-powered land and air vehicles are even worse.)

Let's now compare the Zinsser Effect with the Biefeld-Brown Effect, looking at the details. Brown reports in his 1929 article that there are effects on plants and animals, as well as effects from the sun, moon and even slightly from some of the planetary positions. Zinsser also reports beneficial effects on plants and humans, including what he called "bacteriostasis and cytostasis."[9] Brown also refers to the "endogravitic" and "exogravitic" times that were representative of the charging and discharging times. Once the gravitator was charged, depending upon "its gravitic capacity" any further electrical input had no effect. *This is the same phenomenon that Zinsser witnessed* and both agree that the *pulsed voltage generation* was the main part of the electrogravitic effect.[10] Both Zinsser and Brown worked with dielectrics and capacitor plate transducers to produce the electrogravitic force. Both refer to a high dielectric constant material in between their capacitor plates as the preferred type to best insulate the charge. However, Zinsser never experimented with different dielectrics nor higher voltage to increase his force production. This was always a source of frustration for him but he wanted to keep working with water as his dielectric.

B. Electrically Charged Torque Pendulum of Erwin Saxl

Brown particularly worked with a torque (torsion) pendulum arrangement to measure the force production. He also refers the planetary effects being most pronounced *when aligned with the gravitator* instead of perpendicular to it. He compares these results to Saxl and Allen, who worked with an electrically charged torque pendulum.[11] Dr. Erwin Saxl used high voltage in the range of +/- 5000 volts on his very massive torque pendulum.[12] The changes in period of oscillation measurements with solar or lunar eclipses, showed great sensitivity to the shielding effects of

172

gravity during an alignment of astronomical bodies, helping to corroborate Brown's observation in his 1929 article. The pendulum Saxl used was over 100 kilograms in mass.[13] Most interesting were the "unexpected phenomena" which Saxl reported in his 1964 *Nature* article (see ref. 10). The positively charged pendulum had the longest period of oscillation compared to the negatively charged or grounded pendulum. Dirunal and seasonal variations were found in the effect of voltage on the pendulum, with the most pronounced occurring during a solar or lunar eclipse. In my opinion, this demonstrates the basic principles of electrogravitics: high voltage and mass together will cause unbalanced forces to occur. In this case, the electrogravitic interaction was measurable by oscillating the mass of a charged torque pendulum (producing current) whose period is normally proportional to its mass.

C. Electrogravitic Woodward-Nordtvedt Effect

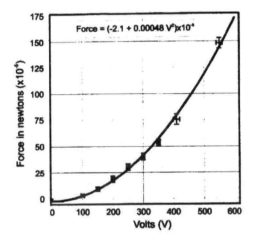
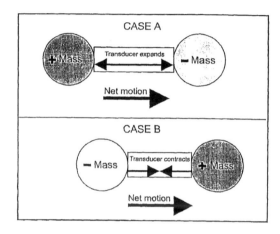

Figure 1. Force Output Vs. Capacitor Voltage Input of a Woodward Force Transducer (Mahood, 2000) and the Net Motion Direction of Cases A and B (Woodward, 2000). *Reported data graph of the Woodward-Nordtvedt effect. Note that the reported force is Newtons ($\times 10^{-5}$) which equals dynes)*

Referring to mass, it is sometimes not clear whether gravitational mass or inertial mass is being affected. The possibility of altering the equivalence principle (which equates the two), has been pursued diligently by Dr. James Woodward[14] (patent cover sheets in Volume II). His prediction, based on Sciama's formulation of Mach's Principle in the framework of general relativity, is that "in the presence of *energy flow*, the inertial mass of an object may undergo sizable variations, changing as the 2nd time derivative of the energy."[15] Woodward, however, indicates that it is the "active gravitational mass" which is being affected but the equivalence principle causes both "passive" inertial and gravitational masses to fluctuate.[16] With barium titanate dielectric between disk capacitors. a 3 kV signal was applied in the experiments of Woodward and Cramer resulting in symmetrical mass fluctuations on the order of centigrams.[17] Cramer actually uses the phrase "Woodward effect" in his AIAA paper, though it is well-known that Nordtvedt was the first to predict noticeable mass shifts in accelerated objects.[18]

The interesting observation which can be made, in light of previous sections, is that Woodward's experimental apparatus *resembles a combination of Saxl's torsion pendulum and Brown's electrogravitic dielectric capacitors.* The differences arise in the precise timing of the pulsed power generation and with input voltage. Recently, 0.01 μF capacitors (Model KD 1653) are being used, in the 50 kHz range (lower than Zinsser's 100 kHz) with the voltage still below 3 kV. Significantly, the thrust or unidirectional force (UDF) is exponential, depending on the square of the applied voltage.[19] However, the micronewton level of force that is produced *is actually the same order of magnitude which Zinsser produced,* who reported his results in dynes (1 dyne = 10^{-5} Newtons).[20] Zinsser had *activators* with masses between 200 g and 500 g and force production of "100 dynes to over one pound."[21] Recently, Woodward has been referring to his transducers as "flux capacitors" (like the movie, *Back to the Future*).[22]

III. Jefimenko's Electrokinetics Explains Electrogravitics

Known for his extensive work with atmospheric electricity, electrostatic motors and electrets, Dr. Oleg Jefimenko deserves significant credit for presenting a valuable theory of the *electrokinetic field*, as he calls it.[23] A W.V. University professor and physics purist at heart, he describes this field as the *dragging force* that electrons exert *on neighboring electric charges*, which is what he says Faraday noted in 1831, when experimenting with parallel wires: a momentary current in the same direction when the current is turned on and then a reverse current in the adjacent wire when the current is turned off.

He identifies the *electrokinetic field* by the vector $\mathbf{E}_k$ where

$$\mathbf{E}_k = -\frac{1}{4\pi\varepsilon_o c^2}\int \frac{1}{r}\left[\frac{\partial \mathbf{J}}{\partial t}\right]dv' \tag{1}$$

It is one of three terms for the electric field in terms of current and charge density. Equations like $\mathbf{F} = q\mathbf{E}$ also apply for calculating force. The significance of $\mathbf{E}_k$, as seen in Eq. 1, is that the electrokinetic field simply the third term of a classical solution for *the electric field* in Maxwell's equations:

$$\mathbf{E} = \frac{1}{4\pi\varepsilon_o}\int\left\{\frac{\rho}{r^2}+\frac{1}{rc}\frac{\partial\rho}{\partial t}\right\}\mathbf{r}\,dv'+\mathbf{E}_k \tag{2}$$

This three-term equation is a causal equation, according to Jefimenko, because it links the electric field $\mathbf{E}$ back the electric charge and its motion (current) which induces it. (He also proves that $\mathbf{E}$ cannot be a causal consequence of a time-variable magnetic field $\partial\mathbf{B}/\partial t$ but instead occurs simultaneously.) This is the essence of electromagnetic induction, *as Maxwell intended*, which is measured by, not caused by, a changing magnetic field. The third electric field term, designated as the electrokinetic field, *is directed along the current direction or parallel to it*. It also exists only as long as the current is changing in time. Lenz' Law is also built into the minus sign. Parallel conductors will produce the strongest induced current.

The significance of Eq. 3 is that the magnetic vector potential is seen to be created by the time integral which amounts to an *electrokinetic impulse* "produced by this current at that point *when the current is switched on*" according to Jefimenko.[24] Of course, a time-varying sinusoidal current will also qualify for production of an electrokinetic field and the vector potential. An important consequence of Eq. 1 is that *the faster the rates of change of current, the larger will be the electrokinetic force*. Therefore, high voltage pulsed inputs are favored.

However, its significance is much more general. "This field can exist anywhere in space and can *manifest itself as a pure force* by its action on free electric charges." All that is required for a measurable force *from a single conductor* is that the change in current density (time derivative) happens very fast (the c^2 in the denominator is also equal to $1/\mu_o\varepsilon_o$ unless the medium has non-vacuum permeability or permittivity).

The electrogravitics experiments of Brown and Zinsser involve a dielectric medium for greater efficacy and charge density. The electrokinetic force on the electric charges (electrons) of the dielectric, according to Eq. (1), is in the *opposite direction of the increasing positive current* (taking into account the minus sign). For parallel plate capacitors, Jefimenko explains that *the strongest induced field is produced between the plates* and so another equation evolves.

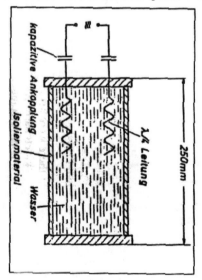

Figure 2. Sample capacitor probe used by Zinsser. *Note the quarter λ/4 wavelength electrodes that indicate an electrically resonant circuit design.*

IV.   Electrokinetic Force Predicts Propulsion Direction

Can Jefimenko's electrokinetic force empirically and qualitatively predict the correct *direction* of the electrogravitic force seen in the Zinsser, Brown, Woodward as well as the yet-to-be-discussed Campbell, Serrano, and Norton AFB craft demonstrations? The following four sections offer empirical evidence for a "prediction" of a force production direction.

1) Starting with *Zinsser's probe diagram* (Fig. 2) from Prof. Peschka's article, it is purposely put on its end in order to compare it with an equivalent parallel plate capacitor (the plates are x distance apart) from Jefimenko's book:[25] Professor Jefimenko performs a calculation of the electrokinetic force in the space between two current-carrying capacitor plates powered by an alternating current. He designates X for the space between the plates where W is the width of each plate and the height is not labeled. His example matches the Zinsser force transducer quite closely.

We note that the current is presumed to be the same in each plate but in opposite directions because it is alternating. Using $E = - \partial A/\partial t$, Jefimenko calculates the electrokinetic field, for the AC parallel plate capacitor with current going in opposite directions, as

$$E_k = -\mu_o \frac{\partial I}{\partial t} \frac{x}{w} \, \mathbf{j} \tag{3}$$

where **j** is the unit vector for the y-axis direction . It is clearly seen that the y-axis points upward in Fig. 3 and so with the minus sign of Eq. 3, the electrokinetic force for the AC parallel plate capacitor *will point downward*. Since Zinsser had his torsion balance on display in Toronto in 1981, I was privileged to verify the direction of the force that is created with his quarter-wave plates oriented as they are in Fig. 2. The torsion balance is built so that the capacitor probe can only be deflected *downward* from the horizontal. *The electrokinetic force is in the same direction.*

2) Looking at *Brown's electrogravitic force direction* from Fig. 3 in his 1929 article "How I Control Gravitation,"

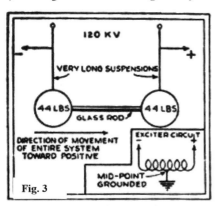

Fig. 3

A SIMPLE TYPE OF GRAVITATOR IS SHOWN IN THE ABOVE ILLUSTRATION.

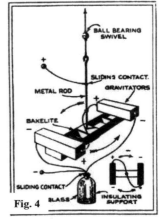

Fig. 4

A GRAVITATOR ROTOR IS SIMPLY AN ASSEMBLY OF UNITS SO MADE THAT ROTATION RESULTS UNTIL THE IMPULSE IS EXHAUSTED.

we see that the positive lead is on the right side of the picture. Also, the arrow below *points to the right* with the caption, "Direction of movement of entire system toward positive." Examining the electrokinetic force of Eq. 1 in this article, we note that the increasing positive current comes in by convention in the positive lead and points to the left. Therefore, considering the minus sign, the direction of the electrokinetic force will be *to the right*. Checking with Fig. 4 of the 1929 Brown article, the same *confirmation of induced electrokinetic force direction.*[26] Thus, with Zinsser's and Brown's gravitators, *the electrokinetic theory provides a useful explanation and it is accurate for prediction of the resulting force direction.*

It is also worthwhile noting that T.T. Brown also indicates in that article,

"when the direct current with high voltage (75 – 300 kilovolts) is applied, the gravitator swings up the arc ... but it does not remain there. The pendulum then gradually returns to the vertical or starting position, even while the potential is maintained...Less than five seconds is required for the test pendulum to reach the maximum amplitude of the swing, but from thirty to eighty seconds are required for it to return to zero."

The Anti-Gravity Files

This phenomenon is *remarkably the same type of response that Zinsser recorded* with his experimental probes. Jefimenko's theory helps explain the rapid response, since the change of current happens in the beginning. However, the slow discharge in both experiments (which Zinsser called a "storage effect") needs more consideration. Considering the electrokinetic force of Eq. 3 and the +/- derivative, we know that the slow draining of a charged capacitor, most clearly seen in Fig. 1 of Brown's 1929 article, will produce a decreasing current out of the + terminal (to the right) and in Eq. 3, this means the derivative is negative. Therefore, *the slow draining of current will produce a weakening electrokinetic force* but *in the same direction as before*! The force will thus sustain itself to the right during discharge.

3) It is reasonable at this stage to also suggest that the electrokinetic theory will also predict the direction of *Woodward's UDF* but instantaneous analysis needs to be made to compare current direction into the commercial disk capacitors and the electrokinetic force on the dielectric charges. In every electrogravitics or electrokinetics case, it can be argued, the "neighboring charges" to a capacitor plate will necessarily be those in the dielectric material, which are <u>polarized.</u> The bound electron-lattice interaction *will drag the lattice material with them*, under the influence of the electrokinetic force. If the combination of physical electron acceleration (which also can be regarded as current flow) and the AC signal current flow can be resolved, it may be concluded that an instantaneous electrokinetic force, depending on dI/dt, contributes to the Woodward-Nordtvedt effect.

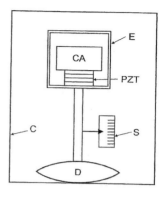

Figure 5. Woodward's #6,098,924 patented impulse engine, also called a "flux capacitor." *The PZT provides nanometer-sized movements that are timed to an AC signal input. A torsion balance has been used*

4) The *Campbell and Serrano capacitor modules* seen in their patented drawings in Figs. 6 and 7, as well as the *Electrogravitic Craft Demonstration unit (Norton AFB, 1988),*[27] can also be analyzed with the electrokinetic force, in the same way that the Brown gravitator force was explained in paragraph (2) above. The current flows in one direction through the capacitor-dielectric and the force is produced in the opposite direction. The Norton AFB electrogravitic craft just has bigger plates with radial sections but the current flow still occurs at the center, *across the plates*. The Serrano patent diagram is also very similar in construction and operation. Campbell's NASA patents include #6,317,310, #6,411,493, and #6,775,123.

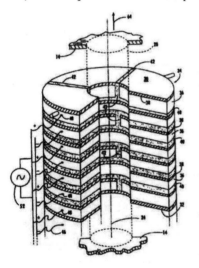

Figure 7. Capacitor propulsion device. *alternating metal and dielectric layers from Serrano's PCT patent WO 00/58623 with upward thrust direction indicated and + and – polarity designated on the side.*

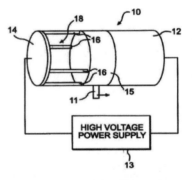

Figure 6. Capacitor module from Campbell's NASA patent #6,317,310 which creates a thrust force. Disk 14 is copper; Struts 16 are dielectrics; Cylinder 15 is a dielectric; Cylinder 12 is an axial capacitor plate; Support post 11 is also dielectric.

V. Electrokinetic Theory Observations

For parallel plate capacitor impulse probes, like Zinsser, Serrano, Campbell, the Norton AFB craft and both of Brown's models, the electrokinetic field of Eq. 3 provides a working model that seems to predict the *nature and direction of the force during charging and discharging phases.* More detailed information is needed for each example in order to actually calculate the theoretical electrokinetic force and compare it with experiment. We note that Eq. 3 also does not suffer the handicap of Eq. 1 since no c^2 term occurs in the denominator. Therefore, it can be concluded that AC fields operating on parallel plate capacitors should create *significantly larger* electrogravitic forces than other geometries with the same dI/dt. However, the current I is usually designated as $I_o\sin(\omega t)$ and its derivative is a sinusoid as well. Therefore, a detailed analysis is needed for each specific circuit and signal to determine the outcome.

Eq. 3 also seems to suggest a *possible enhancement* of the force *if a permeable dielectric (magnetizable) is used.* Then, the value for μ of the material would normally be substituted for μ_o.[28]

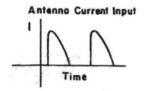

Antenna Current Input

I

Time

Figure 8. A possible electrokinetic force current waveform. *Schlicher propulsion patent #5,142,861*

A further observation of both Eq. 1 and Eq. 3 is that very fast changes in current, such as *a current surge or spark discharge* has to produce the most dynamic electrokinetic force, since dI/dt will be very large.[29] *The declining current surge,* or the negatively sloped dI/dt however, should create an opposing force until the current reverses direction. *Creative waveshaping seems to be the answer* to this obvious dilemma. Fortunately, a few similar inventions use pulse power electric current generators to create propulsion. The Taylor patent #5,197,279 "Electromagnetic Energy Propulsion Engine" uses huge currents to produce magnetic field repulsion. The Schlicher patent #5,142,861 "Nonlinear Electromagnetic Propulsion System and Method" predicts hundreds of pounds of thrust with tens of kiloamperes input. The Schlicher antenna current input is a rectified current surge produced with an SCR-triggered DC power source (see Fig. 8). The resulting waveform has a very steep leading edge but a *slowly declining trailing edge,* which should also be desirable for the electrokinetic force effect.[30] Furthermore, if this waveform is continued into the negative current direction below the horizontal axis, <u>all</u> of that region reinforces the electrokinetic force, with no opposite forces. Therefore, *a complete sinusoidal wave,* with Schlicher-style steep rise-times is recommended for a signal that contributes to a unidirectional force during 75% of its cycle.

Another observation that should be mentioned is that this electrokinetic force theory does not include the mass contribution to the electrogravitic force which Saxl, Woodward, and Brown's 1929 gravitator emphasize. A contributor to *Electrogravitics II,* Takaaki Musha offers a derived equation for electrogravitics *that does include a mass term* but not a derivative term. His model is based on the charge displacement or "deformation" of the atom under the influence of a capacitor's 18 kV high voltage field and his experimental results are encouraging. He also includes a reference to Ning Li and her *gravitoelectric theory.*[31]

A final concern, which may arise from the very nature of the electrokinetic force description, is the difficulty of conceptualizing or simply accepting the possibility of an *unbalanced force creation pushing against space.* This author has wrestled with this problem in other arenas for years. Three examples include (1) the homopolar generator which creates *back torque* that ironically, *pushes against space* to implement the Lorentz force to slow down the current-generating spinning disk.[32] Secondly (2), there is the intriguing *spatial angular momentum discovery* by Graham and Lahoz.[33] They have shown, reminiscent of Feynman's "disk paradox," that the vacuum is the seat of Newton's third law. A torsion balance is their chosen apparatus as well to demonstrate the pure reaction force with induction fields. Their reference to Einstein and Laub's papers cites the time derivative of the Poynting vector $\mathbf{S} = \mathbf{E} \times \mathbf{H}$ integrated over all space to preserve Newton's third law. Graham and Lahoz predict that *magnetic flywheels with electrets* will circulate energy to *push against space.* Lastly, for (3), the Taylor and Schlicher inventions push against space with an unbalanced force that is electromagnetic in origin.

A further confirmation of an electromagnetic explanation for the electrokinetic force empirically can be found in the semiconductor integrated circuit industry. Bothra's US patent #6,191,481 describes an electromigration impeding metallization lines and oxide slots that purposely cause "back-flow" (col. 6, line 25-30). The back-flow of electrons literally causes a force that not only stops electromigration, but if large enough, may perhaps be argued to cause a transfer of momentum to the lattice. This is a direction for high amperage pulsed current experiments to consider for a theoretical foundation for the propulsive force production.

At the Utah chapter meeting of the National Space Society in 2006, a military contractor also described his work with asymmetric capacitors which were summarized as "I levitated a hockey puck" with pulsed currents.

VI. Eye Witness Testimony of Advanced Electrogravitics

Sincere gratitude is given to Mark McCandlish, who has suffered personal trauma for publicizing this work, offers us one of the most conclusive rendition of a covert, flat-bottomed saucer hovercraft seen by dozens of invited eye-witnesses, including a Congressman, at Norton Air Force Base in 1988. When I spoke to Dr. Hal Puthoff about Mark's story, shortly after the famous Disclosure Event[34] at the National Press Club in 2001, he explained to me that he had already performed due diligence on it and checked on each individual to verify the details of the story. Hal explains,

> "All I was able to determine by my due diligence was: (1) to independently interview the source of the story and verify that, indeed he did tell the story to the individual who had passed it on to me, and (2) to independently interview yet another individual who had heard a similar story from a separate source. BUT, I was never able to verify that the story itself was true, only that there were two individuals who said it was true. I then corrected you with my statement (exact quote): '... the story remains in my 'gray basket' only as 'possibly' true.'"

Since Dr. Puthoff used to work for the CIA for ten years as a director of Project Stargate, this was quite an endorsement, even if only cautiously optimistic. In analyzing the Electrogravitic Craft Demonstration unit (Norton AFB 1988) diagrammed in Fig. 9, it can be compared to Campbell's and Serrano's patented design. A lot can be learned from studying the intricacies of this advanced design, including the use of a distributor cap style of pulse discharge and multiple symmetric, radial plates with dielectrics in between. (See reference 27 for Mark's details.) It also remains in my 'gray basket' as possibly true.

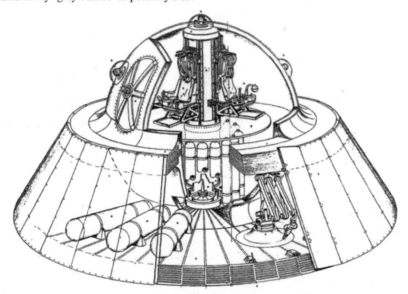

Figure 9.   Electrogravitic Craft Demonstration Unit (Norton AFB, 1988)   - courtesy of Mark McCandlish

Today, we still use World War II technology on land and in space. My sincere hope is that the validating science contained in *Electrogravitics II* will accelerate the civilian adaptation of this propulsion technology.

[1] Valone, Thomas, *Electrogravitics Systems Volume I: Reports on a New Propulsion Methodology*, 6th edition, Integrity Research Institute, Maryland, 2008, ISBN 978-0-9641070-0-7. http://www.integrityresearchinstitute.org/electrogravitics.html

[2] Loder, T., "Outside the Box Space Propulsion and Energy Technology for the 21st Century" AIAA-2002-1131

[3] Valone, Thomas, *Electrogravitics II: Validating Reports on a New Propulsion Methodology*, 3rd edition, 2008, p. 71.   URL at http://www.amazon.com/s/ref=nb_ss_gw?url=search-alias%3Daps&field-keywords=electrogravitics+

[4] Zinsser, R.G. "Mechanical Energy from Anisotropic Gravitational Fields" First Int'l Symp. on Non-Conventional Energy Tech. (FISONCET), Toronto, 1981. Proceedings available from PACE, 100 Bronson Ave #1001, Ottawa, Ontario K1R 6G8

[5] Valone, Thomas *The Zinsser Effect: Cumulative Electrogravity Invention of Rudolf G. Zinsser*, Integrity Research Institute, 2005, 130 pages, ISBN 0-9641070-2-3

[6] Cravens, D.L. "Electric Propulsion/Antigravity" *Electric Spacecraft Journal*, Issue 13, 1994, p. 30 http://www.electricspacecraft.com/journal.htm

[7] Peschka, W., "Kinetobaric Effect as Possible Basis for a New Propulsion Principle," *Raumfahrt-Forschung*, Feb, 1974. Translated version appears in *Infinite Energy*, Issue 22, 1998, p. 52 http://www.infinite-energy.com and in *The Zinsser Effect* book.

[8] Valone, Thomas, "Inertial Propulsion: Concept and Experiment, Part 1" Proc. of Inter. Energy Conver. Eng. Conf., 1993, Available as IRI Report #608.

[9] See "Pulsed Electromagnetic Field Health Effects" IRI Report #418 and *Bioelectromagnetic Healing: A Rationale for Its Use* ISBN 978-0-9641070-5-2 book by this author, which explain the beneficial therapy which PEMFs produce on biological cells.

[10] Mark McCandlish's Testimony (p. 131 of *Electrogravitics II*) shows that the Air Force took note in that the electrogravitic demonstration craft shown at Norton AFB in 1988 had a rotating distributor for electrically pulsing sections of multiply-layered dielectric and metal plate pie-shaped sections with high voltage discharges.

[11] See Saxl patent #3,357,253 "Device and Method for Measuring Gravitational and Other Forces" which uses +/- 5000 volts.

[12] Saxl, E.J., "An Electrically Charged Torque Pendulum" *Nature*, July 11, 1964, p. 136

[13] Saxl & Allen, "Observations with a Massive Electrified Torsion Pendulum: Gravity Measurements During Eclipse," IRI Report #702.(Note: 2.2 lb = 1 kg)

[14] Graph of Fig. 1 from online report, Woodward and Mahood, "Mach's Principle, Mass Fluctuations, and Rapid Spacetime Transport," California State University Fullerton, Fullerton CA 92634

[15] Cramer et al., "Tests of Mach's Principle with a Mechanical Oscillator" AIAA-2001-3908 email: cramer@phys.washington.edu

[16] Woodward, James F. "A New Experimental Approach to Mach's Principle and Relativistic Gravitation, *Found. of Phys. Letters*, V. 3, No. 5, 1990, p. 497

[17] Compare Fig. 1 graph to Brown's ONR graph on P.117 of Volume I

[18] Nordtvedt, K. *Inter. Journal of Theoretical Physics*, V. 27, 1988, p. 1395

[19] Mahood, Thomas "Propellantless Propulsion: Recent Experimental Results Exploiting Transient Mass Modification" Proc. of STAIF, 1999, CP458, p. 1014 (Also see Mahood Master's Thesis www.serve.com/mahood/thesis.pdf)

[20] For comparison, 1 Newton = 0.225 pounds

[21] Zinsser, FISONCET, Toronto, 1981, p. 298

[22] Woodward, James "Flux Capacitors and the Origin of Inertia" *Foundations of Physics*, V. 34, 2004, p. 1475. Also see "Tweaking Flux Capacitors" *Proc. of STAIF*, 2005

[23] Jefimenko, Oleg *Causality, Electromagnetic Induction and Gravitation*, Electret Scientific Co., POB 4132, Star City, WV 26504, p. 29

[24] Jefimenko, p. 31

[25] Jefimenko, p. 47

[26] Brown's second patent #2,949,550 (see Patent Section: two electrokinetic saucers on a maypole) has movement toward the positive charge, so the same electrokinetic theory explained above works for both.

[27] McCandlish, Mark, "Testimony of Mr. Mark McCandlish, December 2000," *Electrogravitics II*, Integrity Research Institute, 2005, p. 131

[28] Einstein and Laub, *Annalen der Physik*, V. 26, 1908, p.533 and p. 541 – two articles on the subject of a moving capacitor with a "dielectric body of considerable permeability." Specific equations are derived predicting the resulting EM fields. Translated articles are reprinted in *The Homopolar Handbook* by this author (p. 122-136). Also see Clark's dielectric homopolar generator patent #6,051,905.

[29] Commentary to Eq. 2 states an electrokinetic impulse is produced when the "current is switched on," which implies a very steep leading edge of the current slope.

[30] See the Taylor and Schlicher patents in the Patent Section. – Ed note

[31] Ning Li was the Chair of the 2003 Gravitational Wave Conference. The *CD Proceedings* of the papers is available from http://www.IntegrityResearchInstitute.org

[32] Valone, Thomas, *The Homopolar Handbook: A Definitive Guide to Faraday Disk and N-Machine Technologies*, Integrity Research Institute, Third Edition, 2001. ISBN 0-9641070-1-5 http://www.amazon.com/s/ref=nb_ss_gw?url=search-alias%3Daps&field-keywords=homopolar+handbook

[33] Graham and Lahoz, "Observation of Static Electromagnetic Angular Momentum in vacuo" *Nature*, V. 285, May 15, 1980, p. 129

[34] See the authoritative book by Dr. Steven Greer, *Disclosure: Military and Government Witnesses Reveal the Greatest Secretes in Modern History*, Crossing Point, 2001. It provides the testimony of each witness who participated in the event, plus many more.

179

The Anti-Gravity Files

Chapter 10

APPENDIX

Anti-Gravity
Documents

The Anti-Gravity Files

Proc. Natl. Acad. Sci. USA
Vol. 74, No. 7, pp. 2664–2666, July 1977
Physics

Electromagnetism and gravitation

(electromagnetic fields induced by rotation/ionic crystals/electrets)

EDWARD TELLER

University of California, Lawrence Livermore Laboratory, P.O. Box 808, Livermore, California 94550

Contributed by Edward Teller, July 30, 1976

ABSTRACT   Generation of electric fields in rapidly rotating insulators is discussed and calculated. An interesting effect is expected in TlCl. A possible appearance of magnetic fields near rapidly rotating gravitating bodies is proposed. The simple suggestion made here would lead to magnetic fields of negligible magnitude.

In the last decades of his life, Einstein attempted to construct a unified field theory. Having reduced gravitation to the principles of geometry, he hoped for a similar system that would include gravitation and electromagnetism. Today, after being puzzled by nuclear forces and after having discovered scores of "elementary particles," we know that in a unified field theory much more needs to be unified. It seems that in physics we did not run fast enough to remain in the same place. This paper is an attempt to discuss some possible relations between gravitation and electromagnetism.

We shall make use of the equivalence principle and replace gravity by acceleration. First, the straightforward problem will be discussed of how an acceleration, conveniently available in rotating bodies, will give rise to electromagnetism. The well-known effects due to the orientation of magnets (i.e., orientation of spins) by the rotation will not be reviewed. We shall concentrate on effects due to the acceleration of charged particles.

In the second part of this paper, magnetism that might arise in the vacuum induced by the rotation of a gravitating body (pulsar or black hole) will be considered.

Polarization induced by acceleration

Electric potentials due to centrifugal forces do occur in rotating metals. The idea that the free electrons are crowded toward bigger radii is naive. The main effect of the centrifugal forces is exercised on the positive ions. The density will be lower near the axis of rotation than at bigger radii. The electrons will neutralize the positive ions. However, when the degeneracy of the electron gas is taken into account the electrons are squeezed out of the denser regions. Thus, a small positive charge will appear near the surface of the rotating body (1).

A much bigger effect can be expected in ionic crystals in which the positive and negative ions, having different masses, will be subject to different centrifugal forces. The polarization of a rotating needle made of an alkali halide will actually be quite similar to the polarization caused by an electric field. The difference is that an electric field acts both on the electrons and on the ions, whereas the centrifugal force acts practically only on the heavy ions. This obvious effect seems to have received little attention.

Let us call the relative displacement of the positive and negative ions x, the masses of the two ions M_+ and M_-, the distance from the axis of rotation r, and the number of revo-

lutions per second ω; then the displacement is given by:*

$$qx = \frac{1}{2}(M_+ - M_-)r\omega^2 \qquad [1]$$

where q is the constant of proportionality in the restoring forces that keep the ions in their equilibrium positions.

The polarization P_{rot} due to the rotation will be

$$P_{rot} = nxe_n = (ne_n/2q)(M_+ - M_-)r\omega^2 \qquad [2]$$

where n is the number of ion pairs per cm^3 and e_n is the effective charge of the ions (that is, the dipole moment due to the relative displacement x divided by that displacement). This charge e_n is dependent on the shape of the crystal (2). The subscript n indicates that the crystal we use has the shape of a needle. The polarization P_{rot} can be measured, and this will give an experimental determination of the quantity e_n/q.

Another application of the same principle is to replace in a complex solid one isotope by another one with the mass difference ΔM. The resulting difference of polarizations caused by the rotation ΔP_{rot} will be rather involved since a force on one atomic species will produce displacement in other atoms. The measurement of ΔP_{rot}, together with other properties of the solid, can contribute to the information about effective charges that are analogous to e_n.

A particularly interesting application would be to rotate an electret at the temperature near the transition point. If the crystal has no ordered permanent dipole moment but we are near the phase transition where such a dipole is established, the rotation will produce big effects which go to infinity as the transition temperature is approached. These measurements may turn out to be relatively easy. Furthermore, they may give interesting insight into the nature of the transition. More explicitly, the application to an electret near its transition point will indicate the relative contributions that the various atoms make in the fluctuations from a state with no dipole to a state with a permanent dipole. Such fluctuations may be enhanced by an electric field or by the forces due to rotation. It remains true that the average polarization induced by the rotation will give information concerning the contributions of various atoms. The measurement of polarizations will lead to quantitative structural information only in combination with data on the vibrational spectrum and the position of atoms before and after the phase transition.

---

* The right-hand side of [1] is the part of the centrifugal force that acts in opposite directions on the positive and negative ions. The portion $1/2(M_+ + M_-)r\omega^2$ that acts on them in the same direction gives rise to compression and makes a negligible contribution in ionic crystals to the polarization. In an element that is an insulator, for instance, sulfur, one would expect only a polarization due to the variation of compression along the needle. This case should be investigated for the sake of comparison. It might yield some unexpected results.

Table 1. Table of data substituted in Eq. 12.

Crystal	k*	k_0	ω_{tr}[†] ($\times 10^{-13}$)	ρ[‡]
NaCl	5.90	2.385[§]	3.62	2.165
KCl	4.75	2.22[¶]	2.97	1.984
KBr	4.90	2.46[§]	2.23	2.75
KI	5.6	2.81	1.98	3.13
AgCl	11.2	4.29	2.31	5.56
AgBr	12.2	5.08	1.67	6.473
TlCl	31.9	5.05[¶]	2.06	7.004

* Squares of refractive indices (3).
[†] Calculated from values of λ (4).
[‡] Ref. 3.
[§] Ref. 5.
[¶] Ref. 6.

Cubic lattices of the type A⁺B⁻

We shall now return to the case of the alkali halides and similar ionic lattices. Here the magnitude of the effect can be calculated from known data. This case, therefore, yields no new information but may serve as a test.

The force constant q of Eq. 1 is simply related to the frequency of the residual rays ω_{tr}, the frequency of a vibration whose wavelength is long compared to the lattice distance (but short compared to the dimensions of the crystal) and whose propagation vector is perpendicular to the displacement of the ions.

According to Lyddane et al. (formula 5′ of ref. 2),

$$\frac{M_+M_-}{M_+ + M_-}\omega_{tr}^2 = \frac{4\pi ne_n^2}{k - k_0} \qquad [3]$$

where k is the dielectric constant of the crystal and k_0 is the smaller value that k assumes if the ions are not permitted to move. This latter value, k_0, may be obtained by extrapolating the square of the refractive index from the optical region to long wavelength, with exclusion of the infrared spectrum.

In the same paper the formula

$$q = \omega_{tr}^2(M_+M_-/M_+ + M_-) \qquad [4]$$

is implied.

Since Eq. 3 is not derived in the quoted paper but is stated as a generalization of other results. its derivation will be indicated here. We consider a needle of the alkali halide crystal in an external electric field E_0 parallel to the needle. The dipole e_nx will be induced per ion pair. Equating the force on this ion pair with the restoring force qx, one has

$$e_nE_0 = qx. \qquad [5]$$

One should note that the force constant q depends on the shape of the crystal and the orientation of the displacement within the crystal. (It may have been more consistent to use the notation q_n.)

The polarization of the crystal is connected with the electric field by

$$4\pi P = (k - 1)E_0. \qquad [6]$$

This polarization is composed of the ionic and electronic contributions. The electronic contribution is defined by

$$4\pi P_{el} = (k_0 - 1)E_0 \qquad [7]$$

while the ionic contribution is the sum of the dipole moments due to the ionic displacements

$$4\pi P_{ion} = 4\pi ne_nx. \qquad [8]$$

Table 2. Values for $4\pi P_{rot}$ electric field at tip of rotating needle for $r\omega^2 = 10^8$ cm·sec^{-2} in units of mV/cm

Crystal	$4\pi P_{rot}$
NaCl	−1.77
KCl	0.39
KBr	−4.48
KI	−9.87
AgCl	16.71
AgBr	6.51
TlCl	70.25

From [5], [6], [7], [8], and $P = P_{ion} + P_{el}$

$$q = 4\pi ne_n^2/(k - k_0) \qquad [9]$$

follows.

In a transverse vibration the electric field is parallel to the nodal planes. Let us assume that our needle is thick compared to the wavelength and the nodal planes are parallel to the needle. Then the electric field is continuous on the faces of the needle and, disregarding end-effects near the point of the needle, all formulas given above apply. Eq. 4 holds for the force constant for a needle (with displacement along the needle) as well as for the force constant of a transverse vibration. Eq. 3 follows from [4] and [9].

From [3] and [4], we obtain

$$\frac{e_n}{q} = \left[\frac{(k - k_0)(M_+ + M_-)}{4\pi nM_+M_-\omega_{tr}^2}\right]^{1/2}. \qquad [10]$$

Substituting this into [2] we get the electric field $4\pi P_{rot}$ which should be observed at the tip of the needle rotated around an axis perpendicular to its long dimension

$$4\pi P_{rot} = (M_+ - M_-)\frac{r\omega^2}{\omega_{tr}}\left[\frac{\pi(k - k_0)\, n(M_+ + M_-)}{M_+M_-}\right]^{1/2} \qquad [11]$$

This formula may be simplified by introducing the density $\rho = n(M_+ + M_-)$ of the ionic crystal

$$4\pi P_{rot} = (M_+ - M_-)\frac{r\omega^2}{\omega_{tr}}\left[\frac{\pi\rho(k - k_0)}{M_+M_-}\right]^{1/2}. \qquad [12]$$

(Incidentally, this polarization gives rise to a volume-density of charge within the needle of the value $(M_+ - M_-)$ $\omega^2/4\omega_{tr}[\rho(k - k_0)/\pi M_+M_-]^{1/2}$ with a compensating surface density of charge at the needle. From this the potential in the neighborhood of the needle can be derived.)

As examples we shall consider the alkali halides NaCl, KCl, KBr; and KI and also the analogous compounds AgCl, AgBr, and TlCl. The data that were used for k, k_0, ω_{tr}, and ρ are given in Table 1 (3–6). Values for the electric field $4\pi P_{rot}$ near the tip of the needle are listed in Table 2 for $r\omega^2 = 10^8$ cm·sec^{-2} in units of mV/cm. It seems quite possible to measure these surface fields $E = 4\pi P_{rot}$ because values of $r\omega^2 > 10^9$ cm·sec^{-2} can be easily obtained in the centrifuge. The samples in question will probably not stand these accelerations. Nevertheless, the effects probably can be observed or even measured. The actual limit of observation is about 10^{-3} mV/cm.

One should note the high value for $4\pi P_{rot}$ in the case of TlCl. (Incidentally, the measurements for k, as given by different authors, differ by as much as 50% for this compound.) The corresponding values for e_n exceed by a great amount the charge of an electron. The picture of the ion carrying a given charge is no longer a good approximation. The vibration corresponding to ω_{tr} may induce a gradual change in the nature of the bond. Indeed, the salt may be in the neighborhood of a phase change. For TlCl, the sign given in Table 2 (which is based on the assumption that Tl carries the positive charge) need not be correct. To carry out the experiment on a rotating needle may be, therefore, particularly interesting in this case.

Magnetic fields near massive rotating stars

The general question of an interaction between gravitation and electromagnetism is, of course, far more basic. Indeed, if the vacuum is a potential source of electrons, positrons, and other particles, then one might imagine that in a vacuum a rotating gravitational field may give rise to electromagnetic forces, just as such forces do arise in rotating crystals. Without attempting to construct a theory one may try to find a formula that has the correct dimensions and the right symmetry properties.

Near a rapidly rotating gravitational object one may, for instance, assume that a magnetic field will appear given by[†]

$$H = (e/c^3)[\vec{g} \times \vec{\omega}] \qquad [13]$$

with e equal to the charge of the electrons, $\vec{g}$ the local gravitational acceleration, and $\vec{\omega}$ the angular velocity of the rotating body. The expression $[\vec{g} \times \vec{\omega}]$ is the vector product of $\vec{g}$ and $\vec{\omega}$.

The dimensions of [13] are easily checked. Time reversal inverts the sign of ω but leaves g unchanged. Thus, H changes its sign under this operation, which is the correct behavior. Inversion (replacement of all space coordinates by their negative values) changes g but not ω. Thus, H would change its sign under inversion, which is incorrect. One should remember, however, that inversion should be accompanied by the interchange of particles and antiparticles. This inverts the sign of e and leads to the correct result that H will not change under inversion combined with charge conjugation. It should be remembered that we are looking for an effect in vacuum. Therefore, a priori, electrons and positrons must have an equal role. Beyond these arguments, which show consistency, I can find no physical reason that fields as given in [13] will really exist. It should be noted that this field, even if real, is exceedingly small.

On the surface of a rotating mass [13] will give, of course, a higher value of H than at greater distances. Since, on the surface, the centrifugal acceleration should be less than g, we find

$$\omega < (g/r)^{1/2}. \qquad [14]$$

Therefore, the absolute value of H will be limited by

$$|H| < \frac{e}{r^2}\left[\frac{(gr)^{1/2}}{c}\right]^3 \approx \frac{e}{r^2}\left(\frac{v_{esc}}{c}\right)^3. \qquad [15]$$

Here v_{esc} is the escape velocity on the surface of the rotating object. The magnetic field will be, therefore, numerically less than an electric field due to a single electron located at the center of the rotating body. The factor $(v_{esc}/c)^3$ can become unity only on the surface of a black hole.

It seems that the elementary conditions to satisfy symmetry relations and to give the correct dimensional behavior suggest an extremely weak coupling between gravitation and magnetism. It cannot, of course, be excluded that a big numerical factor should appear in an equation analogous to [13]. But the factor would have to be truly enormous to make the coupling significant.

Conclusion

The first part of this paper deals with a practical matter that does not appear to be novel; the second part deals with an approach that may be novel but does not seem practical. I am reminded of the professor who told his student: "Your thesis contains material which is new and material which is correct. Unfortunately, what is new is not correct and what is correct is not new." Even so, I hope that some may find one or the other part of this discussion interesting.

It is a pleasure to express my indebtedness to my friend Stanley A. Blumberg whose suggestions and questions gave rise to the considerations presented above. I had several stimulating discussions with Prof. Jesse Beams. These discussions made it clear to me that his centrifuge techniques could be used to obtain more information about insulators, specifically TlCl.

1. Schiff, L. I. & Barnhill, M. V., III (1966) *Phys. Rev.* 151, 1067.
2. Lyddane, R. H., Sachs, R. G. & Teller, E. (1941) *Phys. Rev.* 59, 673.
3. Weast, R. C., ed. (1971) *Handbook of Chemistry and Physics* (Chemical Rubber Company, Cleveland, OH), 52nd ed.
4. Gray, D. E., ed. (1963) *American Institute of Physics Handbook* (McGraw-Hill, New York), 2nd ed., pp. 6–126.
5. Laboratory for Insulation Research, MIT (1953). "Technical Report 57," *Tables of Dielectric Materials*, Vol. IV.
6. Laboratory for Insulation Research, MIT (1957) "Technical report 119," *Tables of Dielectric Materials*, Vol. V.

[†] It would be equivalent to assume a vector potential equal to $e\phi\vec{\omega}/c^3$ where ϕ is the gravitational potential.

Experimental findings of Lifters, Asymmetrical Capacitor Thrusters, and similar electrogravitic devices

Francis X. Canning[*]

Simply Sparse® Technologies, 59 Delrose Drive, Morgantown, WV, 26508

Asymmetrical Capacitor Thrusters have been proposed as a source of propulsion. For over eighty years it has been known that a thrust results when a high voltage is placed across an asymmetrical capacitor, when that voltage causes a leakage current to flow. However, there is surprisingly little experimental or theoretical data explaining this effect. This paper reports on the results of tests of several Asymmetrical Capacitor Thrusters (ACTs). The thrust they produce has been measured for various voltages, polarities, and ground configurations and their radiation in the VHF range has been recorded. These tests were performed at atmospheric pressure and at various reduced pressures. A simple model for the thrust was developed. The model assumed the thrust was due to electrostatic forces on the leakage current flowing across the capacitor. It was further assumed that this current involves charged ions which undergo multiple collisions with air. These collisions transfer momentum. All of the measured data was consistent with this model.

Introduction

THIS paper describes experiments that are designed to explain an effect first observed in 1922. A graduate student, T.T Brown, working under his advisor, Dr. Paul Biefeld, noticed a force in a device when a high voltage was applied. This effect is sometimes called the Biefeld-Brown effect. T.T. Brown received a patent in Great Britain for the use of this effect in 1928[1].

More recently, this effect has been used to produce devices commonly called "Lifters." Lifters are generally light devices that have a high voltage supplied by attached wires. They have generated much interest in hobbyists as they lift off of the ground in a way that appears magical to the casual observer. One such device was patented in 1964[2].

A common feature of these devices is that they apply a high voltage to an asymmetrical capacitor. Some of these devices are called Asymmetrical Capacitor Thrusters. Not only are they asymmetrical, but they generally also have sharp edges or sharp corners. One normally does not think of a capacitor as consuming power its charged state. However, the combination of sharp features and high voltage tends to produce a small leakage current. Potentials in the range of 50 to100 thousand volts are commonly used.

Lifters in particular have generated significant popular interest. It is relatively easy to build devices demonstrating the forces produced. A number of explanations have appeared as to the mechanism that produces the force. There is surprisingly little in the peer reviewed literature on this topic. Some mechanisms proposed in non peer reviewed sources would suggest that these devices might work in a vacuum, although we could find no experiments done in the last fifty years that gave any evidence of this.

This report describes some recent experiments[3,4] that were designed to explain some of the confusing lore about how these devices functioned. For example, some reports suggested that they always created a force towards the side of the capacitor with the sharper physical features. Other reports state this was not always true, and the polarity of the applied voltage did not matter, or that the polarity is what determined the direction of the force.

The work that is reviewed here performed experiments on a variety of devices in air and other gasses at atmospheric pressure and at highly reduced pressures. Voltages, currents, VHF electromagnetic interference and the resulting forces were measured. As a result, we were able to determine the features of these devices that increased the force produced. Also, the design of the experiment showed the features that determine the direction of the force produced. The magnitude of the force that was observed was compared to a simple model based on momentum transfer from the ions of the leakage current to the surrounding atmosphere.

---

[*] FXC@IEEE.org, AIAA non member.

I. Design of Lifters and Asymmetrical Capacitor Thrusters

The specific designs that were tested were used because they both generated a relatively strong force and because they had features that would help in determining the mechanism that produced the thrust. There have been several reports on tests of such devices[3-6]. However, this paper concentrates on reviewing the interpretation of the tests reported in[3,4], in which the present author participated. First, the classic design of a lifter will be discussed. Then, the designs that are generally called Asymmetrical Capacitors will be discussed. Different designs have more and less "sharpness", which will be noted and compared against the experimental results in later sections.

A. Lifter Geometries

A typical lifter is made from materials such as aluminum foil and wire. The wire above the aluminum foil (see Figure 1) is on the top side. The wire may be considered as a sharper surface than the edge of the aluminum foil. The two wires are charged at different potentials. For illustration, the top wire is shown as plus, while the other polarity is just as common.

Lifters seem to most often have only such metallic surfaces, although dielectric material may be used. Even larger lifters weigh very little, often less than an ounce.

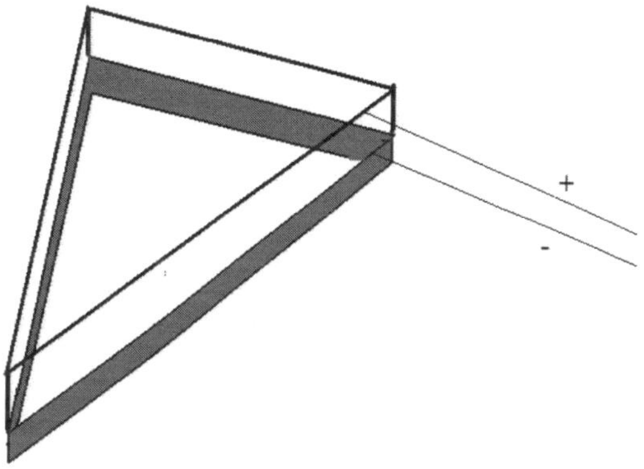

B. Asymmetrical Capacitor Thruster Geometries

Asymmectrical Capacitor Thrusters (ACTs) are similar in design to lifters, but tend to be more recognizable as a capacitor with a strong asymmetry introduced. Nevertheless, one side is more discontinuous than the other. For example, one design uses a disk and a cylinder, where both bodies of revolution share the same axis of rotation.

For a disk and a (hollow) cylinder, the disk is considered to have the sharper features. While

Figure 1. A Typical Lifter

each has a surface with an edge, we believe the presence of the other side of the cylinder softens the discontinuity in the resulting electric fields. A numerical calculation for a two dimensional version of a disk and cylinder was performed in (3) to verify this. It was found that the electric field strength at the edge of the disk, when the cylinder was grounded, was approximately twice the electric field strength at the cylinder when the disk was grounded. This verifies our belief, if the three dimensional case follows the two dimensional case that was simulated by a numerical calculation.

One interesting feature that was tried on some designs consisted of adding individual wires. These wires were obtained from a screen, as one would use in a window. By removing the wires parallel to an edge, only wires pointing out remained there. These created a stronger discontinuity. Thus, some designs that were created for testing had these sharper features and others didn't.

Device 1

Device 4

Figure 2. Two of the Asymmetrical Capacitor Thrusters (ACTs) that were tested.

The Anti-Gravity Files

Four devices were examined in detail. Figure 2 shows the first and fourth devices. Device 2 was similar to Device 1 but it had dielectric material added between the disk and cylinder. Device Three was similar to Device 4, if the wires on the cylinder were removed. Thus, there were two devices with wires and two without.

II. Experimental Setup

Each of the four devices was tested both in a vacuum chamber and in a conducting box with the same dimensions as the vacuum chamber. Two polarities are possible. One has the disk positive and the cylinder negative and the other reverses that polarity.

Fortunately, in building the apparatus it was realized that the ground for the box might be kept at the same potential as either the disk or the cylinder, for either polarity. Thus, there were four combinations of polarity and ground. Observing all four cases proved very illuminating for understanding the physics of these devices. It also explained why previous anecdotal information which had been received appeared to be contradictory. Inconsistent results had been achieved in the past apparently because the location of the ground (i.e. whether it was on the disk or on the cylinder) had not been noted.

III. Qualitative Experimental Results

The results of the experiments in (3) are summarized in A. and B. Part C. gives interpretations of those results.

A. For tests performed in air at atmospheric pressure:

Devices 1 and 2 always produced a force towards the non grounded charged surface.
Devices 3 and 4 always produced a force directed from the cylinder towards the disk.
Devices 1 and 2 produced a larger force when the disk was the non grounded surface.
The polarity (+ Vs -) has only a small effect on the magnitude of the force produced.

When the cylinder was grounded, Devices 3 and 4 produced a significantly larger force than Devices 1 and 2.
When the cylinder was grounded, Device 4 produced more thrust than Device 3.

The current to the live (non grounded) side was always larger than the current from the grounded side to ground.
When the box containing the apparatus was opened, your arm hair stood on end.

B. For tests performed in a vacuum in air and in other gasses:

At pressures such as 300 torr in air, the results were similar to atmospheric pressure but forces were weaker.
Also, similar results were found in Argon and Nitrogen, but with somewhat smaller forces.

In air, the current flowed in bursts, and VHF radiation was observed.
In Argon and Nitrogen, the current did not flow in bursts and the VHF radiation was absent.

In a significant vacuum, with one exception no force was observed, although experimental sensitivity was low.
That exception was a momentary force that occurred at below than one ten thousandth of a torr, when a significant spark (arcing) was observed. This occurred the first time a voltage was applied after the vacuum chamber had been closed and the pressure reduced.

C. Interpretations of these results:

For the tests performed in air, the air became significantly ionized. This allows a current path from the live side to ground that bypasses the grounded side of the ACT. It is possible that the wires on the cylinder of Device 4 reduced the ionization as compared to Device 3, and resulted in the larger force observed. However, that is not clear since tests were not done that controlled the ionization, such as by flushing the air.

The two devices that were somewhat asymmetrical (Devices 1 and 2) produced a force in a direction that was determined by the location of the ground. This is easy to explain as due to the live side having a large voltage gradient at its surface, and the voltage changing abruptly to the ambient value which is near ground. A large voltage gradient gives a large electric field, which causes ionizaton. The ions have the same charge (whether plus or minus) as the nearby ACT surface, so they are repelled by it.

The two devices that were highly asymmetric devices (Devices 3 and 4) always produced a force in a direction that was determined by the asymmetry. That is, the force was always directed from the cylinder towards the disk, regardless of the polarity or the ground location. A reasonable explanation for this is possible since the air was clearly ionized in the box containing the ACT when it was in use. The ionization was displayed in two ways. First, the current into the live side of the ACT was larger than the current out of the grounded side, through the grounding wire. This significant current showed that current was flowing through the air, showing that it must be ionized. Second, the air was found to make the hairs on your arm stand up straight, showing the air was charged. This suggests that near the ACT, the ambient voltage could be different than ground, due to the charges in the air. Thus, there could be a significant voltage gradient between the grounded side of the ACT and the air around it.

The voltage gradient near the grounded side is expected to be significant, even though it likely is smaller than the voltage gradient near the surface of the live (non ground) side of the ACT. Thus, for a large enough asymmetry, the asymmetry would be the determining factor in the net force on the ACT. That is, charged particles may be created on both sides of the ACT, and they would be repelled from their respective nearby side of the ACT. The charged particles near each side produce forces in opposite directions. Thus, it is quite reasonable that for a strong enough asymmetry, the net force is always directed towards the sharper side (e.g. the one with the wires).

The current was found to flow continuously when an ACT was operated in Argon and in Nitrogen, but to flow in bursts in air. This was both measured directly and measured indirectly by observing the VHF radiation that resulted from the bursts of current when the ACT was used in air. These bursts are called Trichel pulses, and are a known phenomena. Since a force was produced in Argon and in Nitrogen, where these bursts do not occur, it appears that the mechanism of the force is not inherently linked to these Trichel pulses.

The only time a noticeable force was created in a strong vacuum occurred the first time a voltage was applied after the chamber had been closed. A significant arcing was associated with this momentary force. Thus, it is possible that some material was removed from one part of the ACT when that arcing occurred. It is reasonable that moisture due to humid air may have deposited on the ACT while the vacuum chamber was open. Thus, this event may have been due to some material (moisture or otherwise) being ejected from the ACT.

IV. Theories Considered Vs Quantitative Experimental Results

It was seen in the previous section that all of the qualitative observations can be explained by a model using a flow of ions. It remains to consider other theories and see if they are plausible. Also, it remains to see if an ion flow model can predict the magnitude of the force created. Several possible theories for the mechanism that creates the force (or thrust) are considered below.

A. Ablative material:
Due to the high voltages and high electric fields that are present, material might be removed from the disc or cylinder during continuous operation. The possibility that continuously ejected material could provide the force was examined in (3). However, since these devices were operated for a large number of hours with no visible change, an upper limit on the amount of mass that might have been removed was easy to find. Using ejection velocities due to thermal effects it was found that any force created by this would be significantly less than that observed, so this effect is not a significant possible mechanism.

B. Electrostatic Forces involving Image Charges:
This concept is simpler to analyze for a lifter than for our test geometry. One might ask if a lifter has charges that interact with image charges due to a ground plane. For example, a concrete floor might have metal reinforcement that produces a current that may be described by image charges. One might assume a perfectly conduction plane under the lifter (this would produce the strongest forces, so more realistic ones would be significantly weaker). The lifter creates approximately a charged dipole, that interacts with its image dipole. A simple calculation shows the resulting force is many orders of magnitude too weak. Thus, this cannot be the mechanism that produces the force. In fact, there is always an attractive force between a charge and its image, so this effect would pull the lifter down rather than make it rise.

C. Polarizing the Vacuum.

The Anti-Gravity Files

This idea is mentioned because it appears on a number of web sites for "hobbyists" building lifters. It is possible that the reason some hobbyists suggest this is that the energy to create an electron positron pair is about a mega electron volt while the energy to move one charge across an ACT is only an order of magnitude less. If one does not think carefully about the physics, this might suggest that one is partially creating electron-positrons out of a vacuum. However, there is a huge difference between the distance across an ACT (centimeters) and sub atomic distances (smaller than an Angstrom). Thus, this could not be the mechanism for lifters and ACTs, since the polarizing force due to the ACT is one hundred million times weaker than it would be if the ACTs work on a charge were performed over atomic distances. The experimental data from Section IV also appears to be inconsistent with this explanation. The various cases that cause forces in different directions or not would be difficult to reconcile with such a model.

D. Ion Drift causing Momentum Transfer to Air.

If charged particles at one side of the ACT are accelerated, move to the other side and decelerate to move with the ACT again, no net force is produced. The net change in momentum of such a particle is zero. For a system consisting of these particles and the ACT, the average force on these particles would be zero so the average force on the ACT would also be zero. The only way a continuous force could be created on an ACT due to these particles would be if these particles transferred momentum to something else. Of course, particles moving through air will have collisions and thus transfer momentum to that air. This is analogous to a propeller moving through air and transferring momentum.

Charged particles would have a large number of collisions traveling from one side of an ACT to the other, and as a result would always be moving much slower than the thermal velocity at sea level and room temperature. Thus, their collision rate would be approximately unchanged due to their motion. With this assumption, it is simple to compute the force produced for a given voltage, distance across the ACT, and current. Further, it may be assumed that all of the charged particles flow in one direction. With these assumptions, the force that would be expected on an ACT was computed in (3). All of the forces measured in (3) were smaller than the result of this computation. However, for Device 4 with the cylinder grounded and the disk positively charged the result was close at 77% of the computed value. This is considered to be exceptional agreement since there are various loss mechanisms that would decrease the force.

V. Conclusion

This paper reviewed the data from some recent tests performed on Asymmetrical Capacitor Thrusters (ACTs). A number of mechanisms were considered for how their thrust is produced. These mechanisms were considered theoretically and in light of test results. Only one mechanism seems plausible, and it relies on standard elementary physics. This model consists of ions drifting from one electrode to the other under electrostatic forces. They collid with air as they move, slowing them down and increasing the time that each contributes to the force. Each collision transferres momentum to the surrounding air, much as a propeller does. This model was found to be consistent with all of the observations that were made. This included how for certain designs the direction of the force changed with which side of the ACT was grounded. It also predicted how for other designs the direction of the force did not change with which side of the ACT was grounded. Furthermore, it predicted the magnitude of the force (thrust) that was measured. This model also predicted that the direction of the thrust was independent of the polarity of the applied voltage. In spite of previous speculation about possible new physical principles being responsible for the thrust produced by ACTs and lifters, we find no evidence to support such a conclusion. On the contrary, a multitude of details about their operation is fully explained by a very simple theory that uses only electrostatic forces and the transfer of momentum by multiple collisions.

References

[1]Brown, T.T. Great Britain Patent 300,311, "A Method of and an Apparatus for Machine for Producing Force or Motion, 1928.

[2]deSeveresky, A.P., U.S. Patent 3,130,945,"Ioncraft," 1964.

[3]Canning, Francis X.,Melcher, Cory, and Winet, Edwin, "Asymmetrical Capacitors for Propulsion," NASA TM CR-2004-213312.

[4]Canning, Francis X., Campbell, Jonathan, Melcher, Cory, Winet, Edwin, and Knudsen, Steven R., "Asymmetrical Capacitors for Propulsion," Proceedings of the 53rd JANNAF Propulsion Conference, Monterey, CA December 2005.

[5]Talley, R. L., "Twenty First Century Propulsion Concept," NASA TM CR-2004-213312.

[6]Bahder, T. B., and Fazi, C., "Force on an Asymmetrical Capacitor—Final Report, Aug.-Dec. 2002," ARL-TR-3005, NTIS Order Number ADA416740.

THE FLYING SAUCER

A SIMPLIFIED EXPLANATION OF THE APPLICATION OF THE BIEFELD-BROWN EFFECT TO THE SOLUTION OF THE PROBLEM OF SPACE NAVIGATION

By

**Mason Rose, Ph.D., President
University for Social Research**

The scientist and layman both encounter a primary difficulty in understanding the Biefeld-Brown effect and its relation to the solution of the flying saucer mystery.

This difficulty lies in the fact that scientist and layman alike think in electromagnetic concepts, whereas the Biefeld-Brown effect relates to electrogravitation.

Neither scientist nor layman can be expected to know the details of electrogravitation, inasmuch as it is a comparatively recent and unpublicized development. Townsend Brown is the discoverer of electrogravitational coupling.

To date, Townsend Brown is the only known experimental scientist in this new area of scientific development. Thus anyone who wishes to understand electrogravitation and its application to astronautics must be prepared to lay aside the commonly known principles of electromagnetics in order to grasp the essentially different principles of electrogravitation. Electrogravitational effects do not obey the known principles of electromagnetism. Electrogravitation must be understood as an entirely new field of scientific investigation and technical development.

Perhaps the most efficient method of inducing an understanding of electrogravitation is to review the evolutionary development of electromagnetism.

From the smallest atom to the largest galaxy, the universe operates on three basic forces—namely, electricity, magnetism and gravitation. These three forces can be represented as follows: *(BUT SHOULD THEY BE ?)*

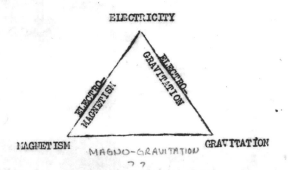

Taken separately, neither is of much practical use. Electricity by itself is static electricity and therefore functionless. It will make your hair stand on end, but that is about all.

Magnetism by itself has few practical applications aside from the magnetic compass, thereas gravity simply keeps objects and people pinned to the earth.

However, when these are coupled to work in combination with each other, almost endless technical applications arise. To date, our total electrical development is based on the coupling of electricity with magnetism, which provides the basis for the countless uses we make of electricity in modern societies.

Farraday conducted the first productive empirical experiments with electromagnetism around 1830, and Maxwell did the basic theoretical work in 1865.

The application of electromagnetism to microscopic and submicroscopic particles was accomplished by Max Planck's work in quantum physics about 1890; and then in 1905 Einstein came forward with relativity, which dealt with gravitation as applied to celestial bodies and universal mechanics.

It is principally out of the work of these four great scientists that our electrical developments ranging from the simple light bulb to the complexities of nuclear phy sics have emerged.

In 1923 Professor Biefeld of Denison University suggested to his protege, Townsend Brown, certain experiments which led to the discovery of the Biefeld-Brown effect and, ultimately , to the electrogravitational energy spectrum. After 28 years of investigation by Brown into this coupling effect between electricity and gravitation, it appears that for each electromagnetic phenomenon there exists an electrogravitational analogue. This means, from the technical and commercial viewpoint, potentialities for future development and exploitation as great or greater than the present electrical industry. When one considers that electromagnetism is basic to the telephone, telegraph, radio, television, radar, electric generators and motors, power production and distribution, and is an indispensible adjunct to transportation of all kinds, one can see that the possibility of a parallel, but different, development in electrogravitation has almost unlimited prospects.

The first empirical experiments conducted by Townsend Brown had the characteristic of simplicity which has marked most other great scientific advancements. These concerned the behavior of a condenser when charged with electricity .

The first startling revelation was that if placed in free suspension with the poles horizontal, the condenser, when charged, exhibited a forward thrust toward the positive pole. A reversal of polarity caused a reversal of the direction of thrust. The experiment was set up as follows:

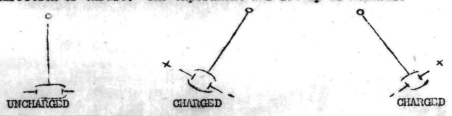

UNCHARGED   CHARGED   CHARGED

The antigravity effect of vertical thrust is demonstrated by balancing a condenser on a beam balance and then charging it. After charging, if the positive pole is pointed upward, the condenser moves up.

If the charge is reversed and the positive pole pointed downward, the condenser thrusts down. The experiment is conducted as follows:

| UNCHARGED | CHARGED | CHARGED |

These two simple experiments demonstrate what is now known as the Biefeld-Brown effect. It is the first and, to the best of our knowledge, the only method of affecting a gravitational field by electrical means. It contains the seeds of control of gravity by men. The intensity of the effect is determined by five factors, which are:

1. The separation of the plates of the condenser---the closer the plates the greater the effect;

2. The ability of the material between the plates to store electrical energy in the form of elastic stress. A measure of this ability is called the "K" of the material. The higher the K the greater the Biefeld-Brown effect.

3. The area of the plates---the greater area giving the greater effect.

4. The voltage difference between the plates---more voltage, more effect.

5. The mass of the material between the plates---the greater the mass, the greater the effect.

It is this fifth point which is inexplicable from the electromagnetic viewpoint and which provides the connection with gravitation.

On the basis of further experimental work from 1923 to 1926, Townsend Brown in 1926 described what he called a "space car." This was a revolutionary method of terrestrial and extraterrestrial flight presented for experiment while motor-propelled planes were yet in a primitive stage.

This engineering feat by Townsend Brown was all the more remarkable when we consider such a machine produces thrust with no moving parts, does not use any aerodynamic principles of flight, and has neither control surfaces nor a propeller. Townsend Brown had discovered the secret of how the flying saucers fly years before any such objects were reported.

Now that the basic differences between electromagnetism and electrogravity have been indicated and the basic principle of the Biefeld-Brown effect has been outlined, we are finally ready to understand the principles of astronautics or the conquest of space.

-4-

The earth creates and is surrounded with a gravitational field which approaches zero as we go far into space. This field presses objects and people to the earth's surface; hence it presses a saucer object to the earth.

However, through the utilization of the Biefeld-Brown effect, the fly - ing saucer can generate an electrogravitational field of its own which modifies the earth's field.

This field acts like a wave, with the negative pole at the top of the wave and the positive pole at the bottom. The saucer travels like a surf-board on the incline of a wave that is kept continually moving by the sau-cer's electrogravitational generator.

Since the orientation of the field can be controlled, the saucer can thus travel on its own continuously generated wave in any desired angle or direction of flight.

The method of controlling the flight of the saucer is illustrated by the following simple diagrams showing the charge variations necessary to ac-complish all directions of flight.

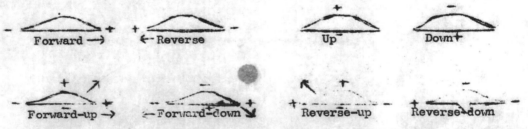

Since the saucer always moves toward its positive pole, the control of the saucer is accomplished simply by varying the orientation of the posi-tive charge. Control, therefore, is gained by switching charges rather than by control surface. Since the saucer is traveling on the incline of a con-tinually moving wave which it generates to modify the earth's gravitational field, no mechanical propulsion is necessary.

Once we understand that the horizontal and vertical controls are ob-tained by shifting the positive pole which turns the field, then we are in a position to extrapolate a finished saucer design.

The top view would be as follows:

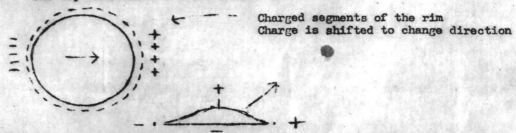

Charged segments of the rim
Charge is shifted to change direction

Upper plate charged positive, lower negative, for lift resultant direction between thrust and lift indicated by arrow.

—5—

The saucer's edge would contain a number of conductor segments, and the saucer would turn in any direction simply by shifting the positive and negative charges to appropriate positions along its edge.

The vertical thrust would be regulated by varying the positive charge on top of the saucer, the amount of thrust being regulated by the amount of charge generated.

Flying saucers in all probability do not utilize external controls for direction, nor do they have any visible means of propulsion. The flying saucers fly on an entirely new principle namely, the Biefeld-Brown electrogravitational effect—and hence do not utilize any of the standard aerodynamic principles of an airfoil. Flying saucers cannot be understood from the traditional principles of aeronautical engineering.

To understand the flying saucers an individual must temporarily bypass these points of view to learn about a new principle—the Biefeld-Brown electrogravitational effect—and then return to the older points of view for critical theoretical analysis and empirical testing.

Years ago, long before saucers as such were reported by observers, Townsend Brown developed a captive flying saucer—that is, a scale model saucer with a free bearing going around a stationary pole.

Brown did not start with round objects—in fact, the first object that flew was a triangle (1), the next a square (2), then a square with the edges cut off (3), and finally a round shaped saucer (4).

The evolutionary development could be graphically expressed as follows:

Since experiments proved the saucer shape most effective, the changes were made for empirical reasons.

Having solved the problem of horizontal thrust, Townsend Brown developed a profile shape which would be most efficient to shape the electrogravitational field for maximum vertical thrust. The final profile that developed was the shape illustrated here:

-6-

The first report of a disc-shaped object in the sky date's back to the sixteenth century. At long intervals during the centuries since have come other reports. Most of them are undoubtedly unreliable as observations, distorted by telling and retelling. But in these older reports, as well as in the very numerous series which has accumulated since 1947, there is a teasing common thread concerning appearance and behavior which makes any certainties about the unreality of flying saucers very insecure.

One of the great difficulties in substantiation of these reports is that, in both appearance and behavior, these objects seem to be simple scientific impossibilities. Here are some of the reasons advanced by technical men to prove the impossibility of devices such as the reports describe:

1. The reports reveal, in most cases, no method of propoulsion which can be understood. There are no propellers in any of the reports. Some reports describe a long flame jet trailing behind a cigar-shaped object. But this flame is orange-red in color, indicating an inefficient combustion which would make it ineffective as a reaction jet such as propels rockets and jet planes. No other known physical laws seemed capable of explaining the observed motion of the objects.

2. The reports describe a range of speed and acceleration from stationary hovering to speeds greater than present-day rockets can deliver. And the changes of rate of motion, the accelerations, are far beyond the capacities of any known man-made vehicles. Flight experts point out that such accelerations would impose impossible stresses on any human or human-like occupants. Therefore, they say, the reports must be false or erroneous.

3. Many of the reports concern night sightings and describe a glow, usually of blue or violet color, around the periphery of the objects. Physicists have noted that such a glow is characteristic of a very high voltage electrical discharge, but add that this suggests no means of explaining the appearance or behavior of the objects described in the reports.

4. The description of shapes and performance seems to indicate a complete or almost complete disregard of aerodynamic principles. The objects seem not to need the support of air as a plane does, nor to depend on the lift provided by properly designed surfaces moved rapidly through an air medium.

These are weighty arguments, PROVIDED THE ASSUMPTIONS BEHIND THEM ARE CORRECT. But now comes physicist Townsend Brown, who has spent the last 28 years exploring the consequences of a simple experiment he performed at the suggestions of Dr. Biefeld in 1923. Dr. Biefeld, professor of physics and astromony at Denison University , former classmate of Einstein in Switzerland, wondered if an electrical condenser, hung by a thread, would have any tendency to move when it was given a heavy electrical charge. Townsend Brown provided the answer. There is such a tendency. But the attempt to understand and explain this motion has occupied him ever since and led to discoveries of truly basic importance.

The observed motion of a charged condenser has been labelled the Biefeld-Brown effect. Studying this effect, Brown pointed out in 1923 that this tendency of a charged condenser to move might easily grow into a new and

basically different method of propulsion. By 1926 he has had described a
"space-car" utilizing this new principle. By 1928 he had built working
models of a boat propelled in this manner. By 1938 he had shown that his
specially designed condensers not only moved but had certain interesting
effects on plants and animals.

All of this, while very exciting, is for most of us just a repetition
of the story scientific development so characteristic of our age. But now
comes the unexpected. Townsend Brown, working in his laboratory, building
models and trying endless variations of size, shape and design of his charged
condensers, made a flying saucer which flew around a maypole BEFORE FLYING
SAUCERS BECAME A NEWSPAPER TOPIC. AND THE REASONS LISTED ABOVE WHICH LED THE
SPECIALISTS TO REJECT THE REPORTS ON OBSERVED SAUCERS PROVED TO BE BOTH EX-
PLICABLE AND NECESSARY TO THEIR OPERATION UNDER THE ELECTROGRAVITATIONAL
PRINCIPLE! *WOW!*

Let us look at out four main objections in a new light.

1. No understood method of propulsion. The saucers made by Brown have
no propellers, no jets, no moving parts at all. They create a modification of the gravitational field around themselves, which is
analagous to putting them on the incline of a hill. They act like
a surfboard on a wave. The surfboard moves without propellers or
jets too, but it is confined to the direction and speed of the water
wave. The electrogravitational saucer creates its own "hill," which
is a local distortion of the gravitational field. Then it takes this
"hill" with it in any chosen direction and at any rate.

2. The second objection concerned the tremendous accelerations which,
on the basis of previous technology, would subject any animal occupants to unbearable stresses. But, says Brown, the occupants of
one of his saucers would feel no stress at all, no matter how sharp
the turn or how great the acceleration. This is because the ship and
the occupants and the load are all responding equally to the wave-
like distortion of the local gravitational field. In an airplane the
propeller pumps air backward and by reaction itself moves forward.
The reaction thrust on the propeller is transferred to the frame of
the aircraft. This frame then shoves the load and occupants forward CONTRARY TO THEIR NATURAL TENDENCY TO MOVE AT A CONSTANT RATE
IN A CONSTANT DIRECTION. But in the saucer no such transfers of
thrust from one member to another occurs. The entire assembly moves
in unison in response to the locally modified gravitational field.
The nearest analogy in our experience is going down in an elevator.
When the elevator starts down, it is not necessary for the elevator
to shove on our bodies---both elevator and passengers share a gravitational tendency to move down. They do so without any shoving or
any stresses between elevator and passengers.

3. Townsend Brown's saucers require a highly charged leading edge, the
positive pole. But such a charged edge produces an electrical corona.
In the largest models made, this develops a decided bluish-violet glow
easily visible in darkness or a dim light. A full-scale ship operating on this principle would be expected to produce a spectacular
corona effect visible for many miles.

4. The outlines and shape of Brown's saucers were the result of electrogravitational considerations---not the result of wind-tunnel tests of
aerodynamic designs. For they move, not on the lift of air, but on

—8—

the lift of a modified gravitational field. In operating saucers such aerodynamic considerations would have to be taken into account to reduce drag and friction, but not to produce lift and thrust.

5. And, finally, when Brown turned his attention to improved ways of generating high voltages, the most promising new method involved the use of a flame jet to convey negative charges astern. This flame was relatively inefficient as a generator if it was adjusted for the best combustion of the fuel. But if it was adjusted to an orange-red color, indicating incomplete combustion of fuel, it conveyed the charges very effectively and set up the required negative space charge behind the ship.

The reasons advanced by the experts to "explain away" the saucer reports, when seen from a new and different viewpoint, appear to be the specific reasons why they can operate—on electrogravitational rather than electromagnetic principles.

The next opinion which must be corrected is the idea of overly intensified supersonic vibration. The Townsend Brown experiments indicate that the positive field which is traveling in front of the saucer acts as a buffer wing which starts moving the air out of the way. This immaterial electrogravitational field acts as an entering wedge which softens the supersonic barrier, thus allowing the material leading edge of the saucer to enter into a softened pressure area. Diagrammed, this would be illustrated as follows:

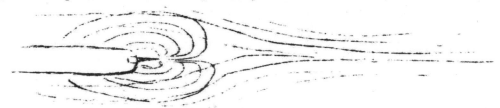

The University for Social Research is ready to offer this experimental finding to the jet airplane and guided missile industry as a practical method of softening the supersonic barrier.

It should be noted that in a jet plane or guided missile the extra weight added to create the Biefeld-Brown electrogravitational effect would be compensated for by the added thrust created by the movement of the plane toward the positive field created in front of its leading edge.

As we have previously stated, for every known electromagnetic effect there is an analogous electrogravitational effect but electrogravitational application and results differ from those of electromagnetic. This presupposes that an entire new electrogravitational industry comparable in size to the present electromagnetic industry will emerge from the theoretical formulations and empirical experiments of Townsend Brown.

The University for Social Research, in presenting the Biefeld-Brown electrogravitational effect, offers to the world new vistas of increased production, betterment of human living and additional economic stability to all countries.

Mason Rose, Ph.D., President
University for Social Research

1312 North Stanley
Hollywood 46, California

April 8, 1952

The Anti-Gravity Report
Updated in 2005

TABLE OF CONTENTS

Integrity Research Institute
5020 Sunnyside Avenue, Suite 209
Beltsville MD 20705

ANCIENT ALIENS ON MARS
By Mike Bara

Bara brings us this lavishly illustrated volume on alien structures on Mars. Was there once a vast, technologically advanced civilization on Mars, and did it leave evidence of its existence behind for humans to find eons later? Did these advanced extraterrestrial visitors vanish in a solar system wide cataclysm of their own making, only to make their way to Earth and start anew? Was Mars once as lush and green as the Earth, and teeming with life? Chapters include: War of the Worlds; The Mars Tidal Model; The Death of Mars; Cydonia and the Face on Mars; The Monuments of Mars; The Search for Life on Mars; The True Colors of Mars and The Pathfinder Sphinx; more. Color section.
252 Pages. 6x9 Paperback. Illustrated. $19.95. Code: AMAR

ANCIENT ALIENS ON THE MOON
By Mike Bara
What did NASA find in their explorations of the solar system that they may have kept from the general public? How ancient really are these ruins on the Moon? Using official NASA and Russian photos of the Moon, Bara looks at vast cityscapes and domes in the Sinus Medii region as well as glass domes in the Crisium region. Bara also takes a detailed look at the mission of Apollo 17 and the case that this was a salvage mission, primarily concerned with investigating an opening into a massive hexagonal ruin near the landing site. Chapters include: The History of Lunar Anomalies; The Early 20th Century; Sinus Medii; To the Moon Alice!; Mare Crisium; Yes, Virginia, We Really Went to the Moon; Apollo 17; more. Tons of photos of the Moon examined for possible structures and other anomalies. 8-Page Color Section.
248 Pages. 6x9 Paperback. Illustrated. $19.95. Code: AAOM

PRODIGAL GENIUS
The Life of Nikola Tesla
by John J. O'Neill
This special edition of O'Neill's book has many rare photographs of Tesla and his most advanced inventions. Tesla's eccentric personality gives his life story a strange romantic quality. He made his first million before he was forty, yet gave up his royalties in a gesture of friendship, and died almost in poverty. Tesla could see an invention in 3-D, from every angle, within his mind, before it was built; how he refused to accept the Nobel Prize; his friendships with Mark Twain, George Westinghouse and competition with Thomas Edison. Tesla is revealed as a figure of genius whose influence on the world reaches into the far future. Deluxe, illustrated edition.
408 pages. 6x9 Paperback. Illustrated. Bibliography. $18.95. Code: PRG

THE CRYSTAL SKULLS
Astonishing Portals to Man's Past
by David Hatcher Childress and Stephen S. Mehler
Childress introduces the technology and lore of crystals, and then plunges into the turbulent times of the Mexican Revolution form the backdrop for the rollicking adventures of Ambrose Bierce, the renowned journalist who went missing in the jungles in 1913, and F.A. Mitchell-Hedges, the notorious adventurer who emerged from the jungles with the most famous of the crystal skulls. Mehler shares his extensive knowledge of and experience with crystal skulls. Having been involved in the field since the 1980s, he has personally examined many of the most influential skulls, and has worked with the leaders in crystal skull research, including the inimitable Nick Nocerino, who developed a meticulous methodology for the purpose of examining the skulls.
294 pages. 6x9 Paperback. Illustrated. Bibliography. $18.95. Code: CRSK

SAUCERS OF THE ILLUMINATI
by Jim Keith, Foreword by Kenn Thomas
Seeking the truth behind stories of alien invasion, secret underground bases, and the secret plans of the New World Order, *Saucers of the Illuminati* offers ground breaking research, uncovering clues to the nature of UFOs and to forces even more sinister: the secret cabal behind planetary control! Includes mind control, saucer abductions, the MJ-12 documents, cattle mutilations, government anti-gravity testing, the Sirius Connection, science fiction author Philip K. Dick and his efforts to expose the Illuminati, plus more from veteran conspiracy and UFO author Keith. Conspiracy expert Keith's final book on UFOs and the highly secret group that manufactures them and uses them for their own purposes: the control and manipulation of the population of planet Earth.
148 Pages. 6x9 Paperback. Illustrated. $12.95. Code: SOIL

DEATH ON MARS
The Discovery of a Planetary Nuclear Massacre
By John E. Brandenburg, Ph.D.

New proof of a nuclear catastrophe on Mars! In an epic story of discovery, strong evidence is presented for a dead civilization on Mars and the shocking reason for its demise: an ancient planetary-scale nuclear massacre leaving isotopic traces of vast explosions that endure to our present age. The story told by a wide range of Mars data is now clear. Mars was once Earth-like in climate, with an ocean and rivers, and for a long period became home to both plant and animal life, including a humanoid civilization. Then, for unfathomable reasons, a massive thermo-nuclear explosion ravaged the centers of the Martian civilization and destroyed the biosphere of the planet. But the story does not end there. This tragedy may explain Fermi's Paradox, the fact that the cosmos, seemingly so fertile and with so many planets suitable for life, is as silent as a graveyard.

278 Pages. 6x9 Paperback. Illustrated. Bibliography. Color Section. $19.95. Code: DOM

BEYOND EINSTEIN'S UNIFIED FIELD
Gravity and Electro-Magnetism Redefined
By John Brandenburg, Ph.D.

Brandenburg reveals the GEM Unification Theory that proves the mathematical and physical interrelation of the forces of gravity and electromagnetism! Brandenburg describes control of space-time geometry through electromagnetism, and states that faster-than-light travel will be possible in the future. Anti-gravity through electromagnetism is possible, which upholds the basic "flying saucer" design utilizing "The Tesla Vortex." Chapters include: Squaring the Circle, Einstein's Final Triumph; A Book of Numbers and Forms; Kepler, Newton and the Sun King; Magnus and Electra; Atoms of Light; Einstein's Glory, Relativity; The Aurora; Tesla's Vortex and the Cliffs of Zeno; The Hidden 5th Dimension; The GEM Unification Theory; Anti-Gravity and Human Flight; The New GEM Cosmos; more. Includes an 8-page color section.

312 Pages. 6x9 Paperback. Illustrated. $18.95. Code: BEUF

THE COSMIC WAR
Interplanetary Warfare, Modern Physics, and Ancient Texts
By Joseph P. Farrell

There is ample evidence across our solar system of catastrophic events. The asteroid belt may be the remains of an exploded planet! The known planets are scarred from incredible impacts, and teeter in their orbits due to causes heretofore inadequately explained. Included: The history of the Exploded Planet hypothesis, and what mechanism can actually explode a planet. The role of plasma cosmology, plasma physics and scalar physics. The ancient texts telling of such destructions: from Sumeria (Tiamat's destruction by Marduk), Egypt (Edfu and the Mars connections), Greece (Saturn's role in the War of the Titans) and the ancient Americas.

436 Pages. 6x9 Paperback. Illustrated.. $18.95. Code: COSW

THE GRID OF THE GODS
The Aftermath of the Cosmic War & the Physics of the Pyramid Peoples
By Joseph P. Farrell with Scott D. de Hart

Farrell looks at Ashlars and Engineering; Anomalies at the Temples of Angkor; The Ancient Prime Meridian: Giza; Transmitters, Nazis and Geomancy; the Lithium-7 Mystery; Nazi Transmitters and the Earth Grid; The Master Plan of a Hidden Elite; Moving and Immoveable Stones; Uncountable Stones and Stones of the Giants and Gods; Gateway Traditions; The Grid and the Ancient Elite; Finding the Center of the Land; The Ancient Catastrophe, the Very High Civilization, and the Post-Catastrophe Elite; Tiahuanaco and the Puma Punkhu Paradox: Ancient Machining; The Black Brotherhood and Blood Sacrifices; The Gears of Giza: the Center of the Machine; Alchemical Cosmology and Quantum Mechanics in Stone; tons more.

436 Pages. 6x9 Paperback. Illustrated. $19.95. Code: GOG

THE SS BROTHERHOOD OF THE BELL
The Nazis' Incredible Secret Technology
by Joseph P. Farrell

In 1945, a mysterious Nazi secret weapons project code-named "The Bell" left its underground bunker in lower Silesia, along with all its project documentation, and a four-star SS general named Hans Kammler. Taken aboard a massive six engine Junkers 390 ultra-long range aircraft, "The Bell," Kammler, and all project records disappeared completely, along with the gigantic aircraft. It is thought to have flown to America or Argentina. What was "The Bell"? What new physics might the Nazis have discovered with it? How far did the Nazis go after the war to protect the advanced energy technology that it represented?

456 pages. 6x9 Paperback. Illustrated. $16.95. Code: SSBB

THE ANTI-GRAVITY HANDBOOK
edited by David Hatcher Childress, with Nikola Tesla, T.B. Paulicki, Bruce Cathie, Albert Einstein and others

The new expanded compilation of material on Anti-Gravity, Free Energy, Flying Saucer Propulsion, UFOs, Suppressed Technology, NASA Cover-ups and more. Highly illustrated with patents, technical illustrations and photos. This revised and expanded edition has more material, including photos of Area 51, Nevada, the government's secret testing facility. This classic on weird science is back in a 90s format!

- **How to build a flying saucer.**
- **Arthur C. Clarke on Anti-Gravity.**
- **Crystals and their role in levitation.**
- **Secret government research and development.**

230 PAGES. 7x10 PAPERBACK. ILLUSTRATED. $16.95. CODE: AGH

ANTI–GRAVITY & THE WORLD GRID

Is the earth surrounded by an intricate electromagnetic grid network offering free energy? This compilation of material on ley lines and world power points contains chapters on the geography, mathematics, and light harmonics of the earth grid. Learn the purpose of ley lines and ancient megalithic structures located on the grid. Discover how the grid made the Philadelphia Experiment possible. Explore the Coral Castle and many other mysteries, including acoustic levitation, Tesla Shields and scalar wave weaponry. Browse through the section on anti-gravity patents, and research resources.

274 PAGES. 7x10 PAPERBACK. ILLUSTRATED. $14.95. CODE: AGW

ANTI–GRAVITY & THE UNIFIED FIELD
edited by David Hatcher Childress

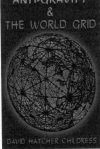

Is Einstein's Unified Field Theory the answer to all of our energy problems? Explored in this compilation of material is how gravity, electricity and magnetism manifest from a unified field around us. Why artificial gravity is possible; secrets of UFO propulsion; free energy; Nikola Tesla and anti-gravity airships of the 20s and 30s; flying saucers as superconducting whirls of plasma; anti-mass generators; vortex propulsion; suppressed technology; government cover-ups; gravitational pulse drive; spacecraft & more.

240 PAGES. 7x10 PAPERBACK. ILLUSTRATED. $14.95. CODE: AGU

THE MYSTERY OF THE OLMECS
by David Hatcher Childress

Lost Cities author Childress takes us deep into Mexico and Central America in search of the mysterious Olmecs, North America's early, advanced civilization. The Olmecs, now sometimes called Proto-Mayans, were not acknowledged to have existed as a civilization until an international archeological meeting in Mexico City in 1942. At this time, the megalithic statues, large structures, ceramics and other artifacts were acknowledged to come from this hitherto unknown culture that pre-dated all other cultures of Central America. But who were the Olmecs? Where did they come from? What happened to them? How sophisticated was their culture? How far back in time did it go? Why are many Olmec statues and figurines seemingly of foreign peoples such as Africans, Europeans and Chinese? Is there a link with Atlantis? In this heavily illustrated book, join Childress in search of the lost cites of the Olmecs!

432 Pages. 6x9 Paperback. Illustrated. Bibliography. $20.00. Code: MOLM

PATH OF THE POLE
Cataclysmic Pole Shift Geology
by Charles H. Hapgood

Maps of the Ancient Sea Kings author Hapgood's classic book *Path of the Pole* is back in print! Hapgood researched Antarctica, ancient maps and the geological record to conclude that the Earth's crust has slipped on the inner core many times in the past, changing the position of the pole. *Path of the Pole* discusses the various "pole shifts" in Earth's past, giving evidence for each one, and moves on to possible future pole shifts. Packed with illustrations, this is the sourcebook for many other books on cataclysms and pole shifts.

356 PAGES. 6x9 PAPERBACK. ILLUSTRATED. $16.95. CODE: POP

MAPS OF THE ANCIENT SEA KINGS
Evidence of Advanced Civilization in the Ice Age
by Charles H. Hapgood

Charles Hapgood's classic 1966 book on ancient maps produces concrete evidence of an advanced world-wide civilization existing many thousands of years before ancient Egypt. He has found the evidence in the Piri Reis Map that shows Antarctica, the Hadji Ahmed map, the Oronteus Finaeus and other amazing maps. Hapgood concluded that these maps were made from more ancient maps from the various ancient archives around the world, now lost. Not only were these unknown people more advanced in mapmaking than any people prior to the 18th century, it appears they mapped all the continents. The Americas were mapped thousands of years before Columbus. Antarctica was mapped when its coasts were free of ice!

316 PAGES. 7x10 PAPERBACK. ILLUSTRATED. BIBLIOGRAPHY & INDEX. $22.95. CODE: MASK

THE LAND OF OSIRIS
An Introduction to Khemitology
by Stephen S. Mehler
Was there an advanced prehistoric civilization in ancient Egypt who built the great pyramids and carved the Great Sphinx? Did the pyramids serve as energy devices and not as tombs for kings? Mehler has uncovered an indigenous oral tradition that still exists in Egypt, and has been fortunate to have studied with a living master of this tradition, Abd'El Hakim Awyan. Mehler has also been given permission to present these teachings to the Western world, teachings that unfold a whole new understanding of ancient Egypt . Chapters include: Egyptology and Its Paradigms; Asgat Nefer—The Harmony of Water; Khemit and the Myth of Atlantis; The Extraterrestrial Question; more.
272 PAGES. 6x9 PAPERBACK. ILLUSTRATED. COLOR SECTION. BIBLIOGRAPHY. $18.00 CODE: LOOS

REICH OF THE BLACK SUN
Nazi Secret Weapons and the Cold War Allied Legend
by Joseph P. Farrell
Why were the Allies worried about an atom bomb attack by the Germans in 1944? Why did the Soviets threaten to use poison gas against the Germans? Why did Hitler in 1945 insist that holding Prague could win the war for the Third Reich? Why did US General George Patton's Third Army race for the Skoda works at Pilsen in Czechoslovakia instead of Berlin? Why did the US Army not test the uranium atom bomb it dropped on Hiroshima? Why did the Luftwaffe fly a non-stop round trip mission to within twenty miles of New York City in 1944? *Reich of the Black Sun* takes the reader on a scientific-historical journey in order to answer these questions. Arguing that Nazi Germany actually won the race for the atom bomb in late 1944, *Reich of the Black Sun* then goes on to explore the even more secretive research the Nazis were conducting into the occult, alternative physics and new energy sources. The book concludes with a fresh look at the "Nazi Legend" of the UFO mystery by examining the Roswell Majestic-12 documents and the Kecksburg crash in the light of parallels with some of the super-secret black projects being run by the SS. *Reich of the Black Sun* is must-reading for the researcher interested in alternative history, science, or UFOs!
352 PAGES. 6x9 PAPERBACK. ILLUSTRATED. BIBLIOGRAPHY. $16.95. CODE: ROBS

ANDROMEDA: THE SECRET FILES
The Flying Submarines of the SS
By David Hatcher Childress
Childress brings us the amazing story of the German Andromeda craft, designed and built during WWII. Along with flying discs, the Germans were making long, cylindrical airships that are commonly called motherships—large craft that house several smaller disc craft. It was not until 1989 that a German researcher named Ralf Ettl, living in London, received an anonymous packet of photographs and documents concerning the planning and development of at least three types of unusual craft—including the Andromeda. Chapters include: Gravity's Rainbow; The Motherships; The MJ-12, UFOs and the Korean War; The Strange Case of Reinhold Schmidt; Secret Cities of the Winged Serpent; The Green Fireballs; Submarines That Can Fly; The Breakaway Civilization; more. Includes a 16-page color section.
382 Pages. 6x9 Paperback. Illustrated. $22.00 Code: ASF

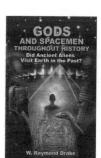

GODS AND SPACEMEN THROUGHOUT HISTORY
Did Ancient Aliens Visit Earth in the Past?
By W. Raymond Drake
From prehistory, flying saucers have been seen in our skies. As mankind sends probes beyond the fringes of our galaxy, we must ask ourselves: "Has all this happened before? Could extraterrestrials have landed on Earth centuries ago?" Drake spent many years digging through huge archives of material, looking for supposed anomalies that could support his scenarios of space aliens impacting human history. Chapters include: Spacemen; The Golden Age; Sons of the Gods; Lemuria; Atlantis; Ancient America; Aztecs and Incas; India; Tibet; China; Japan; Egypt; The Great Pyramid; Babylon; Israel; Greece; Italy; Ancient Rome; Scandinavia; Britain; Saxon Times; Norman Times; The Middle Ages; The Age of Reason; Today; Tomorrow; more.
280 Pages. 6x9 Paperback. Illustrated. $18.95. Code: GSTH

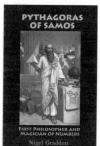

PYTHAGORAS OF SAMOS
First Philosopher and Magician of Numbers
By Nigel Graddon
This comprehensive account comprises both the historical and metaphysical aspects of Pythagoras' philosophy and teachings. In Part 1, the work draws on all known biographical sources as well as key extracts from the esoteric record to paint a fascinating picture of the Master's amazing life and work. Topics covered include the unique circumstances of Pythagoras' birth, his forty-year period of initiations into all the world's ancient mysteries, his remarkable meeting with a physician from the mysterious Etruscan community, Part 2 comprises, for the first time in a publicly available work, a metaphysical interpretation of Pythagoras' Science of Numbers.
294 Pages. 6x9 Paperback. Illustrated. $18.95. Code: PYOS

HITLER'S SUPPRESSED AND STILL-SECRET WEAPONS, SCIENCE AND TECHNOLOGY
by Henry Stevens
In the closing months of WWII the Allies assembled mind-blowing intelligence reports of supermetals, electric guns, and ray weapons able to stop the engines of Allied aircraft—in addition to feared x-ray and laser weaponry. Chapters include: The Kammler Group; German Flying Disc Update; The Electromagnetic Vampire; Liquid Air; Synthetic Blood; German Free Energy Research; German Atomic Tests; The Fuel-Air Bomb; Supermetals; Red Mercury; Means to Stop Engines; more.
335 Pages. 6x9 Paperback. Illustrated. $19.95. Code: HSSW

SECRETS OF THE MYSTERIOUS VALLEY
by Christopher O'Brien
No other region in North America features the variety and intensity of unusual phenomena found in the world's largest alpine valley, the San Luis Valley of Colorado and New Mexico. Since 1989, Christopher O'Brien has documented thousands of high-strange accounts that report UFOs, ghosts, crypto-creatures, cattle mutilations, skinwalkers and sorcerers, along with portal areas, secret underground bases and covert military activity. This mysterious region at the top of North America has a higher incidence of UFO reports than any other area of the continent and is the publicized birthplace of the "cattle mutilation" mystery. Hundreds of animals have been found strangely slain during waves of anomalous aerial craft sightings. Is the government directly involved? Are there underground bases here?
460 pages. 6x9 Paperback. Illustrated. Bibliography. $19.95. Code: SOMV

QUEST FOR ZERO-POINT ENERGY
Engineering Principles for "Free Energy"
by Moray B. King
King expands, with diagrams, on how free energy and anti-gravity are possible. The theories of zero point energy maintain there are tremendous fluctuations of electrical field energy embedded within the fabric of space. King explains the following topics: Tapping the Zero-Point Energy as an Energy Source; Fundamentals of a Zero-Point Energy Technology; Vacuum Energy Vortices; The Super Tube; Charge Clusters: The Basis of Zero-Point Energy Inventions; Vortex Filaments, Torsion Fields and the Zero-Point Energy; Transforming the Planet with a Zero-Point Energy Experiment; Dual Vortex Forms: The Key to a Large Zero-Point Energy Coherence. Packed with diagrams, patents and photos. With power shortages now a daily reality in many parts of the world, this book offers a fresh approach very rarely mentioned in the mainstream media.
224 PAGES. 6x9 PAPERBACK. ILLUSTRATED. $14.95. CODE: QZPE

TAPPING THE ZERO POINT ENERGY
Free Energy & Anti-Gravity in Today's Physics
by Moray B. King
King explains how free energy and anti-gravity are possible. The theories of the zero point energy maintain there are tremendous fluctuations of electrical field energy imbedded within the fabric of space. This book tells how, in the 1930s, inventor T. Henry Moray could produce a fifty kilowatt "free energy" machine; how an electrified plasma vortex creates anti-gravity; how the Pons/Fleischmann "cold fusion" experiment could produce tremendous heat without fusion; and how certain experiments might produce a gravitational anomaly.
180 PAGES. 5x8 PAPERBACK. ILLUSTRATED. $12.95. CODE: TAP

THE FREE-ENERGY DEVICE HANDBOOK
A Compilation of Patents and Reports
by David Hatcher Childress
A large-format compilation of various patents, papers, descriptions and diagrams concerning free-energy devices and systems. *The Free-Energy Device Handbook* is a visual tool for experimenters and researchers into magnetic motors and other "over-unity" devices. With chapters on the Adams Motor, the Hans Coler Generator, cold fusion, superconductors, "N" machines, space-energy generators, Nikola Tesla, T. Townsend Brown, and the latest in free-energy devices. Packed with photos, technical diagrams, patents and fascinating information, this book belongs on every science shelf. With energy and profit being a major political reason for fighting various wars, free-energy devices, if ever allowed to be mass distributed to consumers, could change the world! Get your copy now before the Department of Energy bans this book!
292 PAGES. 8x10 PAPERBACK. ILLUSTRATED. BIBLIOGRAPHY. $16.95. CODE: FEH

WATER REALMS
Ancient Water Technologies and Management
By Karen Mutton

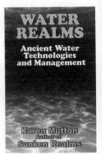

From the flushing toilets of ancient Crete to the qanats of Persia, aqueducts of Rome, cascading tank systems of Sri Lanka and the great baths of the Indus Valley to the eel traps of southern Australia, ancients on all continents were managing water in unique ways. Table of Contents includes: The Minoan Waterworks; Case Study—The Tunnel of Eupalinos, Samos, Sicily; Etruscan Waterworks; Aqueducts; Roman Baths; Case Study—Aqua Sulis; Case Study—The Baths of Caracalla; Flood Control Systems; Hydraulic Works in the Provinces; Case Study—The Pont Du Gard, France; Late Roman & Byzantine Technologies; The Persian Qanat System; Case Study—The Palace of Persepolis; Khmer Empire; Case Study—The Dujiangyan Irrigation System; Hohokam Water Works; Case Study—Teotihuacan; Case Study—The Puquios of Peru; Sardinia Wells; Nymphaea; Celtic Wells; Ancient Fish Traps; more. There are tons of illustrations in this fascinating book!
254 Pages. 6x9 Paperback. Illustrated. $19.95. Code: WTR

THE A.T. FACTOR
A Scientists Encounter with UFOs: Piece For A Jigsaw Part 3
by Leonard Cramp
British aerospace engineer Cramp began much of the scientific anti-gravity and UFO propulsion analysis back in 1955 with his landmark book *Space, Gravity & the Flying Saucer* (out-of-print and rare). His next books (available from Adventures Unlimited) *UFOs & Anti-Gravity: Piece for a Jig-Saw* and *The Cosmic Matrix: Piece for a Jig-Saw Part 2* began Cramp's in depth look into gravity control, free-energy, and the interlocking web of energy that pervades the universe. In this final book, Cramp brings to a close his detailed and controversial study of UFOs and Anti-Gravity.
324 PAGES. 6x9 PAPERBACK. ILLUSTRATED. BIBLIOGRAPHY. INDEX. $16.95. CODE: ATF

COSMIC MATRIX
Piece for a Jig-Saw, Part Two
by Leonard G. Cramp
Leonard G. Cramp, a British aerospace engineer, wrote his first book *Space Gravity and the Flying Saucer* in 1954. Cosmic Matrix is the long-awaited sequel to his 1966 book *UFOs & Anti-Gravity: Piece for a Jig-Saw*. Cramp has had a long history of examining UFO phenomena and has concluded that UFOs use the highest possible aeronautic science to move in the way they do. Cramp examines anti-gravity effects and theorizes that this super-science used by the craft—described in detail in the book—can lift mankind into a new level of technology, transportation and understanding of the universe. The book takes a close look at gravity control, time travel, and the interlocking web of energy between all planets in our solar system with Leonard's unique technical diagrams. A fantastic voyage into the present and future!
364 PAGES. 6x9 PAPERBACK. ILLUSTRATED. BIBLIOGRAPHY. $16.00. CODE: CMX

UFOS AND ANTI-GRAVITY
Piece For A Jig-Saw
by Leonard G. Cramp
Leonard G. Cramp's 1966 classic book on flying saucer propulsion and suppressed technology is a highly technical look at the UFO phenomena by a trained scientist. Cramp first introduces the idea of 'anti-gravity' and introduces us to the various theories of gravitation. He then examines the technology necessary to build a flying saucer and examines in great detail the technical aspects of such a craft. Cramp's book is a wealth of material and diagrams on flying saucers, anti-gravity, suppressed technology, G-fields and UFOs. Chapters include Crossroads of Aerodymanics, Aerodynamic Saucers, Limitations of Rocketry, Gravitation and the Ether, Gravitational Spaceships, G-Field Lift Effects, The Bi-Field Theory, VTOL and Hovercraft, Analysis of UFO photos, more.
388 PAGES. 6x9 PAPERBACK. ILLUSTRATED. $19.95. CODE: UAG

THE TESLA PAPERS
Nikola Tesla on Free Energy & Wireless Transmission of Power
by Nikola Tesla, edited by David Hatcher Childress
David Hatcher Childress takes us into the incredible world of Nikola Tesla and his amazing inventions. Tesla's rare article "The Problem of Increasing Human Energy with Special Reference to the Harnessing of the Sun's Energy" is included. This lengthy article was originally published in the June 1900 issue of *The Century Illustrated Monthly Magazine* and it was the outline for Tesla's master blueprint for the world. Tesla's fantastic vision of the future, including wireless power, anti-gravity, free energy and highly advanced solar power. Also included are some of the papers, patents and material collected on Tesla at the Colorado Springs Tesla Symposiums, including papers on: •The Secret History of Wireless Transmission •Tesla and the Magnifying Transmitter •Design and Construction of a Half-Wave Tesla Coil •Electrostatics: A Key to Free Energy •Progress in Zero-Point Energy Research •Electromagnetic Energy from Antennas to Atoms •Tesla's Particle Beam Technology •Fundamental Excitatory Modes of the Earth-Ionosphere Cavity
325 PAGES. 8x10 PAPERBACK. ILLUSTRATED. $16.95. CODE: TTP

THE FANTASTIC INVENTIONS OF NIKOLA TESLA
by Nikola Tesla with additional material by David Hatcher Childress
This book is a readable compendium of patents, diagrams, photos and explanations of the many incredible inventions of the originator of the modern era of electrification. In Tesla's own words are such topics as wireless transmission of power, death rays, and radio-controlled airships. In addition, rare material on German bases in Antarctica and South America, and a secret city built at a remote jungle site in South America by one of Tesla's students, Guglielmo Marconi. Marconi's secret group claims to have built flying saucers in the 1940s and to have gone to Mars in the early 1950s! Incredible photos of these Tesla craft are included. The Ancient Atlantean system of broadcasting energy through a grid system of obelisks and pyramids is discussed, and a fascinating concept comes out of one chapter: that Egyptian engineers had to wear protective metal head-shields while in these power plants, hence the Egyptian Pharoah's head covering as well as the Face on Mars! •His plan to transmit free electricity into the atmosphere. •How electrical devices would work using only small antennas. •Why unlimited power could be utilized anywhere on earth. •How radio and radar technology can be used as death-ray weapons in Star Wars.
342 PAGES. 6x9 PAPERBACK. ILLUSTRATED. $16.95. CODE: FINT

THE ENERGY GRID
Harmonic 695, The Pulse of the Universe
by Captain Bruce Cathie.
This is the breakthrough book that explores the incredible potential of the Energy Grid and the Earth's Unified Field all around us. Cathie's first book, *Harmonic 33*, was published in 1968 when he was a commercial pilot in New Zealand. Since then, Captain Bruce Cathie has been the premier investigator into the amazing potential of the infinite energy that surrounds our planet every microsecond. Cathie investigates the Harmonics of Light and how the Energy Grid is created. In this amazing book are chapters on UFO Propulsion, Nikola Tesla, Unified Equations, the Mysterious Aerials, Pythagoras & the Grid, Nuclear Detonation and the Grid, Maps of the Ancients, an Australian Stonehenge examined, more.
255 PAGES. 6x9 TRADEPAPER. ILLUSTRATED. $15.95. CODE: TEG

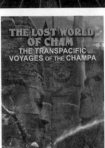

THE LOST WORLD OF CHAM
The Trans-Pacific Voyages of the Champa
By David Hatcher Childress
The mysterious Cham, or Champa, peoples of Southeast Asia formed a megalith-building, seagoing empire that extended into Indonesia, Tonga, and beyond—a transoceanic power that reached Mexico and South America. The Champa maintained many ports in what is today Vietnam, Cambodia, and Indonesia and their ships plied the Indian Ocean and the Pacific, bringing Chinese, African and Indian traders to far off lands, including Olmec ports on the Pacific Coast of Central America. Topics include: Cham and Khem: Egyptian Influence on Cham; The Search for Metals; The Basalt City of Nan Madol; Elephants and Buddhists in North America; The Cham and Lake Titicaca; Easter Island and the Cham; the Magical Technology of the Cham; tons more. 24-page color section.
328 Pages. 6x9 Paperback. Illustrated. $22.00 Code: LPWC

GIANTS ON RECORD
By Jim Vieira and Hugh Newman
Over a 200-year period thousands of newspaper reports, town and county histories, letters, photos, diaries, and scientific journals have documented the existence of an ancient race of giants in North America. Extremely large skeletons ranging from 7 feet up to a staggering 18 feet tall have been reportedly uncovered in prehistoric mounds, burial chambers, caves, geometric earthworks, and ancient battlefields. Strange anatomic anomalies such as double rows of teeth, horned skulls, massive jaws that fit over a modern face, and elongated skulls have also been reported. Color Section.
420 pages. 6x9 Paperback. Illustrated. $19.95. Code: GOR

ADVENTURES OF A HASHISH SMUGGLER
by Henri de Monfreid
Nobleman, writer, adventurer and inspiration for the swashbuckling gun runner in the *Adventures of Tintin*, Henri de Monfreid lived by his own account "a rich, restless, magnificent life" as one of the great travelers of his or any age. The son of a French artist who knew Paul Gaugin as a child, de Monfreid sought his fortune by becoming a collector and merchant of the fabled Persian Gulf pearls. He was then drawn into the shadowy world of arms trading, slavery, smuggling and drugs. Infamous as well as famous, his name is inextricably linked to the Red Sea and the raffish ports between Suez and Aden in the early years of the twentieth century. De Monfreid (1879 to 1974) had a long life of many adventures around the Horn of Africa where he dodged pirates as well as the authorities.
284 Pages. 6x9 Paperback. $16.95. Illustrated. Code AHS

NORTH CAUCASUS DOLMENS
In Search of Wonders
By Boris Loza, Ph.D.
Join Boris Loza as he travels to his ancestral homeland to uncover and explore dolmens firsthand. Throughout this journey, you will discover the often hidden, and surprisingly forbidden, perspective about the mysterious dolmens: their ancient powers of fertility, healing and spiritual connection. Chapters include: Ancient Mystic Megaliths; Who Built the Dolmens?; Why the Dolmens were Built; Asian Connection; Indian Connection; Greek Connection; Olmec and Maya Connection; Sun Worshippers; Dolmens and Archeoastronomy; Location of Dolmen Quarries; Hidden Power of Dolmens; and much more! Tons of Illustrations! A fascinating book of little-seen megaliths. Color section.
252 Pages. 5x9 Paperback. Illustrated. $24.00. Code NCD

LEY LINE & EARTH ENERGIES
An Extraordinary Journey into the Earth's Natural Energy System
by David Cowan & Chris Arnold
The mysterious standing stones, burial grounds and stone circles that lace Europe, the British Isles and other areas have intrigued scientists, writers, artists and travellers through the centuries. How do ley lines work? How did our ancestors use Earth energy to map their sacred sites and burial grounds? How do ghosts and poltergeists interact with Earth energy? How can Earth spirals and black spots affect our health? This exploration shows how natural forces affect our behavior, how they can be used to enhance our health and well being. A fascinating and visual book about subtle Earth energies and how they affect us and the world around them.
368 PAGES. 6x9 PAPERBACK. ILLUSTRATED. BIBLIOGRAPHY. INDEX. $18.95. CODE: LLEE

SECRETS OF THE UNIFIED FIELD
The Philadelphia Experiment, the Nazi Bell, & the Discarded Theory
by Joseph P. Farrell

Farrell examines the discarded Unified Field Theory. American and German wartime scientists determined that, while the theory was incomplete, it could nevertheless be engineered. Chapters include: The Meanings of "Torsion"; The Mistake in Unified Field Theories and Their Discarding by Contemporary Physics; Three Routes to the Doomsday Weapon: Quantum Potential, Torsion, and Vortices; Tesla's Meeting with FDR; Arnold Sommerfeld and Electromagnetic Radar Stealth; Electromagnetic Phase Conjugations, Phase Conjugate Mirrors, and Templates; The Unified Field Theory, the Torsion Tensor, and Igor Witkowski's Idea of the Plasma Focus; tons more.
340 pages. 6x9 Paperback. Illustrated. Bibliography. Index. $18.95. Code: SOUF

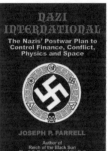

NAZI INTERNATIONAL
The Nazi's Postwar Plan to Control Finance, Conflict, Physics $ Space
by Joseph P. Farrell

Beginning with prewar corporate partnerships in the USA, including some with the Bush family, he moves on to the surrender of Nazi Germany, and evacuation plans of the Germans. He then covers the vast, and still-little-known recreation of Nazi Germany in South America with help of Juan Peron, I.G. Farben and Martin Bormann. Farrell then covers the development and control of new energy technologies including the Bariloche Fusion Project, Dr. Philo Farnsworth's Plasmator, and the work of Dr. Nikolai Kozyrev. Finally, Farrell discusses the Nazi desire to control space, and examines their connection with NASA.
412 pages. 6x9 Paperback. Illustrated. References. $19.95. Code: NZIN

ARKTOS
The Polar Myth in Science, Symbolism & Nazi Survival
by Joscelyn Godwin

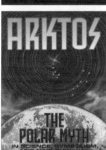

Explored are the many tales of an ancient race said to have lived in the Arctic regions, such as Thule and Hyperborea. Progressing onward, he looks at modern polar legends: including the survival of Hitler, German bases in Antarctica, UFOs, the hollow earth, and the hidden kingdoms of Agartha and Shambala. Chapters include: Prologue in Hyperborea; The Golden Age; The Northern Lights; The Arctic Homeland; The Aryan Myth; The Thule Society; The Black Order; The Hidden Lands; Agartha and the Polaires; Shambhala; The Hole at the Pole; Antarctica; more.
220 Pages. 6x9 Paperback. Illustrated. Bib. Index. $16.95. Code: ARK

THE TIME TRAVEL HANDBOOK
A Manual of Practical Teleportation & Time Travel
edited by David Hatcher Childress

The Time Travel Handbook takes the reader beyond the government experiments and deep into the uncharted territory of early time travellers such as Nikola Tesla and Guglielmo Marconi and their alleged time travel experiments, as well as the Wilson Brothers of EMI and their connection to the Philadelphia Experiment—the U.S. Navy's forays into invisibility, time travel, and teleportation. Childress looks into the claims of time travelling individuals, and investigates the unusual claim that the pyramids on Mars were built in the future and sent back in time. A highly visual, large format book, with patents, photos and schematics. **316 PAGES.**
7X10 PAPERBACK. ILLUSTRATED. $16.95. CODE: TTH

AXIS OF THE WORLD
The Search for the Oldest American Civilization
by Igor Witkowski

Polish author Witkowski's research reveals remnants of a high civilization that was able to exert its influence on almost the entire planet, and did so with full consciousness. Sites around South America show that this was not just one of the places influenced by this culture, but a place where they built their crowning achievements. Easter Island, in the southeastern Pacific, constitutes one of them. The Rongo-Rongo language that developed there points westward to the Indus Valley. Taken together, the facts presented by Witkowski provide a fresh, new proof that an antediluvian, great civilization flourished several millennia ago.
220 pages. 6x9 Paperback. Illustrated. References. $18.95. Code: AXOW

AMERICAN INDIAN MYTHS & MYSTERIES
By Vincent Gaddis

American Indian Myths and Mysteries is veteran fortean Gaddis' fascinating account of the mythology of the Native American—an all-encompassing collection of American Indian legends, truths, and myths. Although much of this ancient heritage has been lost, a great deal has been saved, and there are men and women alive today who remember the lore of their ancestors. Chapters include: Whence Came the Amerind?; Artifacts of a World Forgotten; Mystery of the Megaliths; Tunnels of the Titans; Totem Trails Northward; Columbus Was Late; The European Wanderers; The Lost Colony Enigma; Mystery of the Shaking Tent; Medicine Man Magic; Secrets of the Shamans; The Curse of Tippecanoe; The Great Purification; more.
240 pages. 5x9 Paperback. Illustrated. Bibliography. Index. $16.95 Code: NAM

ANCIENT TECHNOLOGY IN PERU & BOLIVIA
By David Hatcher Childress
Childress speculates on the existence of a sunken city in Lake Titicaca and reveals new evidence that the Sumerians may have arrived in South America 4,000 years ago. He demonstrates that the use of "keystone cuts" with metal clamps poured into them to secure megalithic construction was an advanced technology used all over the world, from the Andes to Egypt, Greece and Southeast Asia. He maintains that only power tools could have made the intricate articulation and drill holes found in extremely hard granite and basalt blocks in Bolivia and Peru, and that the megalith builders had to have had advanced methods for moving and stacking gigantic blocks of stone, some weighing over 100 tons.
340 Pages. 6x9 Paperback. Illustrated.. $19.95 Code: ATP

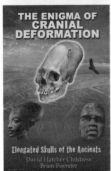

THE ENIGMA OF CRANIAL DEFORMATION
Elongated Skulls of the Ancients
By David Hatcher Childress and Brien Foerster
In a book filled with over a hundred astonishing photos and a color photo section, Childress and Foerster take us to Peru, Bolivia, Egypt, Malta, China, Mexico and other places in search of strange elongated skulls and other cranial deformation. The puzzle of why diverse ancient people—even on remote Pacific Islands—would use head-binding to create elongated heads is mystifying. Where did they even get this idea? Did some people naturally look this way—with long narrow heads? Were they some alien race? Were they an elite race that roamed the entire planet? Why do anthropologists rarely talk about cranial deformation and know so little about it?
250 Pages. 6x9 Paperback. Illustrated. $19.95. Code: ECD

ROSWELL AND THE REICH
By Joseph P. Farrell
Farrell here delves ever deeper into the activities of this nefarious group. In his previous works, Farrell has clearly demonstrated that the Nazis were clandestinely developing new and amazing technologies toward the end of WWII, and that the key scientists involved in these experiments were exported to the Allied countries at the end of the conflict, mainly the United States, in a move called Operation Paperclip. Now, Farrell has meticulously reviewed the best-known Roswell research from UFO-ET advocates and skeptics alike, as well as some little-known source material, and comes to a radically different scenario of what happened in Roswell, New Mexico in July 1947, and why the US military has continued to cover it up to this day. Farrell presents a fascinating case that what crashed may have been representative of an independent postwar Nazi power—an extraterritorial Reich monitoring its old enemy, America, and the continuing development of the very technologies confiscated from Germany at the end of the War.
540 pages. 6x9 Paperback. Illustrated. $19.95. Code: RWR

THE GODS IN THE FIELDS
Michael, Mary and Alice-Guardians of Enchanted Britain
By Nigel Graddon
We learn of Britain's special place in the origins of ancient wisdom and of the "Sun-Men" who taught it to a humanity in its infancy. Aspects of these teachings are found all along the St. Michael ley: at Glastonbury, the location of Merlin and Arthur's Avalon; in the design and layout of the extraordinary Somerset Zodiac of which Glastonbury is a major part; in the amazing stone circles and serpentine avenues at Avebury and nearby Silbury Hill: portals to unimaginable worlds of mystery and enchantment; Chapters include: Michael, Mary and Merlin; England's West Country; The Glastonbury Zodiac; Wiltshire; The Gods in the Fields; Michael, Mary and Alice; East of the Line; Table of Michael and Mary Locations; more.
280 Pages. 6x9 Paperback. Illustrated. $19.95. Code: GIF

GIANTS: MEN OF RENOWN
By Denver Michaels
Michaels runs down the many stories of giants around the world and testifies to the reality of their existence in the past. Chapters and subchapters on: Giants in the Bible; Texts; Tales from the Maya; Stories from the South Pacific; Giants of Ancient America; The Stonish Giants; Mescalero Tales; The Nahullo; Mastodons, Mammoths & Mound Builders; Pawnee Giants; The Si-Te-Cah; Tsul 'Kalu; The Titans & Olympians; The Hyperboreans; European Myths; The Giants of Britain & Ireland; Norse Giants; Myths from the Indian Subcontinent; Daityas, Rakshasas, & More; Jainism: Giants & Inconceivable Lifespans; The Conquistadors Meet the Sons of Anak; Cliff-Dwelling Giants; The Giants of the Channel Islands; Strange Tablets & Other Artifacts; more. Tons of illustrations with an 8-page color section.
320 Pages. 6x9 Paperback. Illustrated. $22.00. Code: GMOR

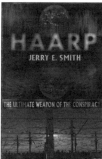

HAARP
The Ultimate Weapon of the Conspiracy
by Jerry Smith

The HAARP project in Alaska is one of the most controversial projects ever undertaken by the U.S. Government. Jerry Smith gives us the history of the HAARP project and explains how works, in technically correct yet easy to understand language. At at worst, HAARP could be the most dangerous device ever created, a futuristic technology that is everything from super-beam weapon to world-wide mind control device. Topics include Over-the-Horizon Radar and HAARP, Mind Control, ELF and HAARP, The Telsa Connection, The Russian Woodpecker, GWEN & HAARP, Earth Penetrating Tomography, Weather Modification, Secret Science of the Conspiracy, more. Includes the complete 1987 Eastlund patent for his pulsed super-weapon that he claims was stolen by the HAARP Project.
256 pages. 6x9 Paperback. Illustrated. Bib. $14.95. Code: HARP

WEATHER WARFARE
The Military's Plan to Draft Mother Nature
by Jerry E. Smith

Weather modification in the form of cloud seeding to increase snow packs in the Sierras or suppress hail over Kansas is now an everyday affair. Underground nuclear tests in Nevada have set off earthquakes. A Russian company has been offering to sell typhoons (hurricanes) on demand since the 1990s. Scientists have been searching for ways to move hurricanes for over fifty years. In the same amount of time we went from the Wright Brothers to Neil Armstrong. Hundreds of environmental and weather modifying technologies have been patented in the United States alone – and hundreds more are being developed in civilian, academic, military and quasi-military laboratories around the world *at this moment!* Numerous ongoing military programs do inject aerosols at high altitude for communications and surveillance operations.
304 Pages. 6x9 Paperback. Illustrated. Bib. $18.95. Code: WWAR

MIND CONTROL, WORLD CONTROL
The Encyclopedia of Mind Control
by Jim Keith

Keith uncovers a surprising amount of information on the technology, experimentation and implementation of Mind Control technology. Various chapters in this shocking book are on early C.I.A. experiments such as Project Artichoke and Project RIC-EDOM, the methodology and technology of implants, Mind Control Assassins and Couriers, various famous "Mind Control" victims such as Sirhan Sirhan and Candy Jones. Also featured in this book are chapters on how Mind Control technology may be linked to some UFO activity and "UFO abductions.
256 Pages. 6x9 Paperback. Illustrated. References. $14.95. Code: MCWC

MIND CONTROL AND UFOS
Casebook on Alternative 3
by Jim Keith

A revised and updated edition of *Casebook on Alternative 3,* Keith's classic investigation of the Alternative 3 scenario as it first appeared on British television over 20 years ago. Keith delves into the bizarre story of Alternative 3, including mind control programs, underground bases not only on the Earth but also on the Moon and Mars, the real origin of the UFO problem, the mysterious deaths of Marconi Electronics employees in Britain during the 1980s, the Russian-American superpower arms race of the 50s, 60s and 70s as a massive hoax, more.
248 Pages. 6x9 Paperback. Illustrated. $14.95. Code: MCUF

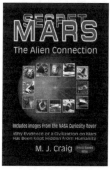

LIQUID CONSPIRACY 2:
The CIA, MI6 & Big Pharma's War on Psychedelics
By Xaviant Haze

Underground author Xaviant Haze looks into the CIA and its use of LSD as a mind control drug; at one point every CIA officer had to take the drug and endure mind control tests and interrogations to see if the drug worked as a "truth serum." Chapters include: The Pioneers of Psychedelia; The United Kingdom Mellows Out: The MI5, MDMA and LSD; LSD becomes Acid; Great Works of Art Inspired and Influenced by Acid; Scapolamine: The CIA's Ultimate Truth Serum; Mind Control, the Death of Music and the Meltdown of the Masses; Big Pharma's War on Psychedelics; The Healing Powers of Psychedelic Medicine; tons more.
240 pages. 6x9 Paperback. Illustrated. $19.95. Code: LQC2

SECRET MARS: The Alien Connection
By M. J. Craig

While scientists spend billions of dollars confirming that microbes live in the Martian soil, people sitting at home on their computers studying the Mars images are making far more astounding discoveries… they have found the possible archaeological remains of an extraterrestrial civilization. Hard to believe? Well, this challenging book invites you to take a look at the astounding pictures yourself and make up your own mind. *Secret Mars* presents over 160 incredible images taken by American and European spacecraft that reveal possible evidence of a civilization that once lived, and may still live, on the planet Mars… powerful evidence that scientists are ignoring! A visual and fascinating book!
352 Pages. 6x9 Paperback. Illustrated. $19.95. Code: SMAR

ARK OF GOD
The Incredible Power of the Ark of the Covenant
By David Hatcher Childress
Childress takes us on an incredible journey in search of the truth about (and science behind) the fantastic biblical artifact known as the Ark of the Covenant. This object made by Moses at Mount Sinai—part wooden-metal box and part golden statue—had the power to create "lightning" to kill people, and also to fly and lead people through the wilderness. The Ark of the Covenant suddenly disappears from the Bible record and what happened to it is not mentioned. Was it hidden in the underground passages of King Solomon's temple and later discovered by the Knights Templar? Was it taken through Egypt to Ethiopia as many Coptic Christians believe? Childress looks into hidden history, astonishing ancient technology, and a 3,000-year-old mystery that continues to fascinate millions of people today. Color section.
420 Pages. 6x9 Paperback. Illustrated. $22.00 Code: AOG

THE TRIANGLE
The Truth Behind the World's Most Enduring Mystery
By Mike Bara
For decades, no single place has intrigued the world more than the baffling Bermuda Triangle. Hundreds of ships, planes and yachts have disappeared in the dark, mysterious waters between Bermuda and Florida. Ships have vanished without a trace only to magically reappear years later in good order but minus their crews, almost as if the intervening years had not even passed—for them. Pilots have reported bizarre problems with their instruments as compasses and guidance systems have spun inexplicably out of control over the shadowy waters of the Triangle. Entire squadrons of military aircraft have disappeared off of radarscopes in clear weather and with no forewarning. Explanations range from alien encounters to rogue waves to twisting unnatural funnel spouts caused by submerged civilizations left over from the days of Atlantis. Find out what really happened to Flight 19, the Navy training flight that last reported "they look like they're from outer space" over the Triangle.
218 Pages. 6x9 Paperback. Illustrated. Notes. $19.95. Code: TRI

LBJ AND THE CONSPIRACY TO KILL KENNEDY
By Joseph P. Farrell
Farrell says that a coalescence of interests in the military industrial complex, the CIA, and Lyndon Baines Johnson's powerful and corrupt political machine in Texas led to the events culminating in the assassination of JFK. Chapters include: Oswald, the FBI, and the CIA: Hoover's Concern of a Second Oswald; Oswald and the Anti-Castro Cubans; The Mafia; Hoover, Johnson, and the Mob; The FBI, the Secret Service, Hoover, and Johnson; The CIA and "Murder Incorporated"; Ruby's Bizarre Behavior; The French Connection and Permindex; Big Oil; The Dead Witnesses: Guy Bannister, Jr., Mary Pinchot Meyer, Rose Cheramie, Dorothy Killgallen, Congressman Hale Boggs; LBJ and the Planning of the Texas Trip; LBJ: A Study in Character, Connections, and Cabals; LBJ and the Aftermath: Accessory After the Fact; The Requirements of Coups D'État; more.
342 Pages. 6x9 Paperback. $19.95 Code: LCKK

THE TESLA PAPERS
Nikola Tesla on Free Energy & Wireless Transmission of Power
by Nikola Tesla, edited by David Hatcher Childress
David Hatcher Childress takes us into the incredible world of Nikola Tesla and his amazing inventions. Tesla's fantastic vision of the future, including wireless power, anti-gravity, free energy and highly advanced solar power. Also included are some of the papers, patents and material collected on Tesla at the Colorado Springs Tesla Symposiums, including papers on: •The Secret History of Wireless Transmission •Tesla and the Magnifying Transmitter •Design and Construction of a Half-Wave Tesla Coil •Electrostatics: A Key to Free Energy •Progress in Zero-Point Energy Research •Electromagnetic Energy from Antennas to Atoms
325 PAGES. 8x10 PAPERBACK. ILLUSTRATED. $16.95. CODE: TTP

THE ENCYCLOPEDIA OF MOON MYSTERIES
Secrets, Conspiracy Theories, Anomalies, Extraterrestrials and More
By Constance Victoria Briggs
Our moon is an enigma. The ancients viewed it as a light to guide them in the darkness, and a god to be worshipped. In modern times, it has been taught that the Moon is simply a dead rock that is caught in Earth's gravity, with no activity. There are stories that suggest that the Moon is home to extraterrestrials, theories that it is not a natural satellite, tales of anomalous lights, and tales that NASA astronauts saw extraterrestrial ships and ruins of an ancient civilization there. There is a rumor that Apollo 13 was saved by extraterrestrials. Another story states that there are extraterrestrial bases on the Moon. Did you know that: Aristotle and Plato wrote about a time when there was no Moon? They even gave a name of an ancient tribe of people that lived during that moonless period; Several of the NASA astronauts reported seeing UFOs while traveling to the Moon?; the Moon might be hollow?; Apollo 10 astronauts heard strange "space music" when traveling on the far side of the Moon? Tons more. Tons of illustrations with A to Z sections for easy reference and reading.
152 Pages. 7x10 Paperback. Illustrated. References. $19.95. Code: EOMM

BIGFOOT NATION
A History of Sasquatch in North America
By David Hatcher Childress

Childress takes a deep look at Bigfoot Nation—the real world of bigfoot around us in the United States and Canada. Whether real or imagined, that bigfoot has made his way into the American psyche cannot be denied. He appears in television commercials, movies, and on roadside billboards. Bigfoot is everywhere, with actors portraying him in variously believable performances and it has become the popular notion that bigfoot is both dangerous and horny. Indeed, bigfoot is out there stalking lovers' lanes and is even more lonely than those frightened teenagers that he sometimes interrupts. Bigfoot, tall and strong as he is, makes a poor leading man in the movies with his awkward personality and typically anti-social behavior. Includes 16-pages of color photos that document Bigfoot Nation!

320 Pages. 6x9 Paperback. Illustrated. References. Color Section. $22.00. Code: BGN

YETIS, SASQUATCH & HAIRY GIANTS
By David Hatcher Childress

Author and adventurer David Hatcher Childress takes the reader on a fantastic journey across the Himalayas to Europe and North America in his quest for Yeti, Sasquatch and Hairy Giants. Childress begins with a discussion of giants and then tells of his own decades-long quest for the Yeti in Nepal, Sikkim, Bhutan and other areas of the Himalayas, and then proceeds to his research into Bigfoot, Sasquatch and Skunk Apes in North America. Chapters include: The Giants of Yore; Giants Among Us; Wildmen and Hairy Giants; The Call of the Yeti; Kanchenjunga Demons; The Yeti of Tibet, Mongolia & Russia; Bigfoot & the Grassman; Sasquatch Rules the Forest; Modern Sasquatch Accounts; more. Inlcudes a 16-page color photo insert.

360 pages. 5x9 Paperback. Illustrated. Bibliography. Index. $18.95. Code: YSHG

THE HOLLOW EARTH
By Dr. Raymond Bernard, introduction by David Hatcher Childress

This classic of 50s & 60s weird UFO literature touches on UFOs from Antarctica, free energy, tunnels in South America, government cover-ups and, of course, the notion that the earth is hollow. "Revealed! The Underground World of Supermen Discovered Under the North Pole." So promised the mysterious Dr. Raymond Bernard to the readers of his book. The front cover said that Dr. Bernard, "noted scholar and author of *The Hollow Earth*, says that the true home of flying saucers is a huge underground world whose entrance is the North Polar opening. Dr. Bernard believes that in the hollow interior of the Earth lives a super-race which wants noting to do with man on the surface.

288 Pages. 5x9 Paperback. Illustrated. $18.95. Code: HERT

THINGS AND MORE THINGS
Myths, Mysteries and Marvels!
by Ivan T. Sanderson

Among the curious "things" are: Flying Saucers; Telepathic Ants; Rocks that Sing—and Kill; Water Monsters; Giant Skulls; Living Dinosaurs; Film of an Abominable Snowman; Frozen Mammoths; Animal ESP; Space Visitors; and much, much more! Absorbing accounts of fantastic phenomena which remain unexplained—and undenied—by science. Chapters include: Globsters; Lake Monsters; Giant Eels; Ringing Rocks; Stone Spheres; Stone Softening; The Toonijuk; UFO Nests; Light Wheels; Flying Rocks; Suspended Animation; Neodinosaurs; Maverick Moas; Giant Skulls; Walkers for Water; Frozen Mammoths; Vile Vortices; Rockets and Rackets; Mechanical Dowsing; tons more.

364 Pages. 6x9 Paperback. Illustrated. Bibliography. Index. $16.95 Code: TMT

ABOMINABLE SNOWMEN: LEGEND COME TO LIFE
The Story of Sub-Humans on Six Continents from the Early Ice Age Until Today
by Ivan T. Sanderson

Do "Abominable Snowmen" exist? Prepare yourself for a shock. In the opinion of one of the world's leading naturalists, not one, but possibly four kinds, still walk the earth! Do they really live on the fringes of the towering Himalayas and the edge of myth-haunted Tibet? From how many areas in the world have factual reports of wild, strange, hairy men emanated? Reports of strange apemen have come in from every continent, except Antarctica.

525 PAGES. 6x9 PAPERBACK. ILLUSTRATED. BIBLIOGRAPHY. INDEX. $19.95. CODE: ABML

INVISIBLE RESIDENTS
The Reality of Underwater UFOS
by Ivan T. Sanderson

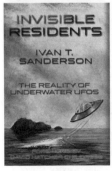

In this book, Sanderson, a renowned zoologist with a keen interest in the paranormal, puts forward the curious theory that "OINTS"—Other Intelligences—live under the Earth's oceans. This underwater, parallel, civilization may be twice as old as Homo sapiens, he proposes, and may have "developed what we call space flight." Sanderson postulates that the OINTS are behind many UFO sightings as well as the mysterious disappearances of aircraft and ships in the Bermuda Triangle. What better place to have an impenetrable base than deep within the oceans of the planet? Sanderson offers here an exhaustive study of USOs (Unidentified Submarine Objects) observed in nearly every part of the world.

298 PAGES. 6x9 PAPERBACK. ILLUSTRATED. BIBLIOGRAPHY. INDEX. $16.95. CODE: INVS

VIMANA:
Flying Machines of the Ancients
by David Hatcher Childress

According to early Sanskrit texts the ancients had several types of airships called vimanas. Like aircraft of today, vimanas were used to fly through the air from city to city; to conduct aerial surveys of uncharted lands; and as delivery vehicles for awesome weapons. David Hatcher Childress, popular *Lost Cities* author and star of the History Channel's long-running show Ancient Aliens, takes us on an astounding investigation into tales of ancient flying machines. In his new book, packed with photos and diagrams, he consults ancient texts and modern stories and presents astonishing evidence that aircraft, similar to the ones we use today, were used thousands of years ago in India, Sumeria, China and other countries. Includes a 24-page color section.

408 Pages. 6x9 Paperback. Illustrated. $22.95. Code: VMA

PIRATES & THE LOST TEMPLAR FLEET
The Secret Naval War Between the Templars & the Vatican
by David Hatcher Childress

Childress takes us into the fascinating world of maverick sea captains who were Knights Templar (and later Scottish Rite Free Masons) who battled the ships that sailed for the Pope. The lost Templar fleet was originally based at La Rochelle in southern France, but fled to the deep fiords of Scotland upon the dissolution of the Order by King Phillip. This banned fleet of ships was later commanded by the St. Clair family of Rosslyn Chapel (birthplace of Free Masonry). St. Clair and his Templars made a voyage to Canada in the year 1298 AD, nearly 100 years before Columbus! Later, this fleet of ships and new ones to come, flew the Skull and Crossbones, the symbol of the Knights Templar.

320 PAGES. 6x9 PAPERBACK. ILLUSTRATED. BIBLIOGRAPHY. $16.95. CODE: PLTF

THE HISTORY OF THE KNIGHTS TEMPLARS
by Charles G. Addison, introduction by David Hatcher Childress

Chapters on the origin of the Templars, their popularity in Europe and their rivalry with the Knights of St. John, later to be known as the Knights of Malta. Detailed information on the activities of the Templars in the Holy Land, and the 1312 AD suppression of the Templars in France and other countries, which culminated in the execution of Jacques de Molay and the continuation of the Knights Templars in England and Scotland; the formation of the society of Knights Templars in London; and the rebuilding of the Temple in 1816. Plus a lengthy intro about the lost Templar fleet and its North American sea routes.

395 PAGES. 6x9 PAPERBACK. ILLUSTRATED. $16.95. CODE: HKT

INSIDE THE GEMSTONE FILE
Howard Hughes, Onassis & JFK
by Kenn Thomas & David Hatcher Childress

Here is the low-down on the most famous underground document ever circulated. Photocopied and distributed for over 20 years, the Gemstone File is the story of Bruce Roberts, the inventor of the synthetic ruby widely used in laser technology today, and his relationship with the Howard Hughes Company and ultimately with Aristotle Onassis, the Mafia, and the CIA. Hughes kidnapped and held a drugged-up prisoner for 10 years; Onassis and his role in the Kennedy Assassination; how the Mafia ran corporate America in the 1960s; more.

320 Pages. 6x9 Paperback. Illustrated. $16.00. Code: IGF

OTTO RAHN AND THE QUEST FOR THE HOLY GRAIL
The Amazing Life of the Real "Indiana Jones"
by Nigel Graddon

Otto Rahn led a life of incredible adventure in southern France in the early 1930s. The Hessian language scholar is said to have found runic Grail tablets in the Pyrenean grottoes, and decoded hidden messages within the medieval Grail masterwork *Parsifal*. The fabulous artifacts identified by Rahn were believed by Himmler to include the Grail Cup, the Spear of Destiny, the Tablets of Moses, the Ark of the Covenant, the Sword and Harp of David, the Sacred Candelabra and the Golden Urn of Manna. Some believe that Rahn was a Nazi guru who wielded immense influence on his elders and "betters" within the Hitler regime, persuading them that the Grail was the Sacred Book of the Aryans, which, once obtained, would justify their extreme political theories and revivify the ancient Germanic myths. But things are never as they seem, and as new facts emerge about Otto Rahn a far more extraordinary story unfolds.

450 pages. 6x9 Paperback. Illustrated. Appendix. Index. $18.95. Code: ORQG

OBELISKS: TOWERS OF POWER
The Mysterious Purpose of Obelisks
By David Hatcher Childress

Childress looks into the enigma of obelisks and their purpose. Egyptologists tell us that obelisks are granite towers that symbolize a ray of the sun—a megalithic symbol of the Sun God Ra, later to be called Aton. Some obelisks weigh over 500 tons and are massive blocks of polished granite that would be extremely difficult to quarry and erect even with modern equipment. Why did ancient civilizations in Egypt, Ethiopia and elsewhere undertake the massive enterprise it would have been to erect a single obelisk, much less dozens of them? Were they energy towers that could receive or transmit energy? With discussions on Tesla's wireless power, and the use of obelisks as gigantic acupuncture needles for earth, Childress shows us what the ancients were trying to achieve with their mysterious obelisks.

320 Pages. 6x9 Paperback. Illustrated. Color Section. $22.00 Code: OBK

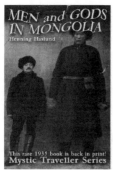

MEN & GODS IN MONGOLIA
by Henning Haslund
Haslund takes us to the lost city of Karakota in the Gobi desert. We meet the Bodgo Gegen, a god-king in Mongolia similar to the Dalai Lama of Tibet. We meet Dambin Jansang, the dreaded warlord of the "Black Gobi." Haslund and companions journey across the Gobi desert by camel caravan; are kidnapped and held for ransom; witness initiation into Shamanic societies; meet reincarnated warlords; and experience the violent birth of "modern" Mongolia.
358 Pages. 6x9 Paperback. Illustrated. $18.95. Code: MGM

TAYOS GOLD
The Archives of Atlantis
by Stan Hall
In 1976, Scottish engineer Hall organized a landmark expedition to Ecuador, involving joint Special Forces and astronaut professor Neil Armstrong as Honorary President and participant. Hall was driven by curiosity about Erich von Däniken's report of a Metal Library allegedly found in the caves by investigator Juan Moricz in the mid-1960s and began an odyssey into the heart of global enigmas: the origins of mankind, Atlantis, Ptolemy's lost city of Cattigara, and the entrance to the Metal Library along the Pastaza River in Ecuador. Chapters include: Juan Moricz-Magyar Extraordinary; Egyptian Tablets of the Mormons; Ecuador: Cradle of Civilization; The Triangle of the Shell, Tunnels Below the Andes; Neil Armstrong: Second Small Step; Into the Tayos Caves; Treasure of the Incas; Explorers Percy Fawcett and George M. Dyott; Valverde's Treasure; Tayos Treasure: Analysis and Location; more.
246 pages. 6x9 Paperback. Illustrated. Bibliography. Appendix. $18.95. Code: TAYG

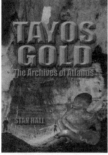

SUNKEN REALMS
A Survey of Underwater Ruins Around the World
By Karen Mutton
Australian researcher Karen Mutton begins by discussing some of the causes for sunken ruins: super-floods; volcanoes; earthquakes at the end of the last great flood; plate tectonics and other theories. From there she launches into a worldwide cataloging of underwater ruins by region. She begins with the many underwater cities in the Mediterranean, and then moves into northern Europe and the North Atlantic. She continues with chapters on the Caribbean and then moves through the extensive sites in the Pacific and Indian Oceans. Comes with plenty of maps, illustrations and rare photos. Places covered in this book include: Tartessos; Cadiz; Morocco; Alexandria; The Bay of Naples; Libya; Phoenician and Egyptian sites; Roman era sites; Yarmuta, Lebanon; Cyprus; Malta; Thule & Hyperborea; Celtic Realms Lyonesse, Ys, and Hy Brasil; Carnac, Brittany; Isle of Wight; Canary and Azore Islands; Bahamas; Cuba; Bermuda; Mexico; Peru; Micronesia; California; Japan; Indian Ocean; Sri Lanka Land Bridge; India; Sumer; Lake Titicaca; and inland lakes in Scotland, Russia, Iran, China, Wisconsin, Florida and more. A unique and fascinating book!
320 Pages. 6x9 Paperback. Illustrated. Bibliography. $20.00. Code: SRLM

LOST CITIES & ANCIENT MYSTERIES OF THE SOUTHWEST
By David Hatcher Childress
Join David as he starts in northern Mexico and searches for the lost mines of the Aztecs. He continues north to west Texas, delving into the mysteries of Big Bend, including mysterious Phoenician tablets discovered there and the strange lights of Marfa. Then into New Mexico where he stumbles upon a hollow mountain with a billion dollars of gold bars hidden deep inside it! In Arizona he investigates tales of Egyptian catacombs in the Grand Canyon, cruises along the Devil's Highway, and tackles the century-old mystery of the Lost Dutchman mine. In Nevada and California Childress checks out the rumors of mummified giants and weird tunnels in Death Valley, plus he searches the Mohave Desert for the mysterious remains of ancient dwellers alongside lakes that dried up tens of thousands of years ago. It's a full-tilt blast down the back roads of the Southwest in search of the weird and wondrous mysteries of the past!
486 Pages. 6x9 Paperback. Illustrated. Bibliography. $19.95. Code: LCSW

SUBTERRANEAN REALMS
By Karen Mutton
Mutton discusses such interesting sites as: Derinkuyu, an underground city in Cappadocia, Turkey that housed 20,000 people; Roman catacombs of Domitilla; Palermo Capuchin catacombs; Alexandria catacombs; Paris catacombs; Maltese hypogeum; Rock-cut structures of Petra; Treasury of Atreus, Mycenae; Elephanta Caves, India; Lalibela, Ethiopia; Tarquinia Etruscan necropolis; Hallstatt salt mine; Beijing air raid shelters; Japanese high command Okinawa tunnels; more. There are tons of illustrations in this fascinating book!
336 Pages. 6x9 Paperback. Illustrated. $19.95. Code: SUBR

LOST CITIES & ANCIENT MYSTERIES OF SOUTH AMERICA
by David Hatcher Childress

Rogue adventurer and maverick archaeologist David Hatcher Childress takes the reader on unforgettable journeys deep into deadly jungles, high up on windswept mountains and across scorching deserts in search of lost civilizations and ancient mysteries. Travel with David and explore stone cities high in mountain forests and hear fantastic tales of Inca treasure, living dinosaurs, and a mysterious tunnel system. Whether he is hopping freight trains, searching for secret cities, or just dealing with the daily problems of food, money, and romance, the author keeps the reader spellbound. Includes both early and current maps, photos, and illustrations, and plenty of advice for the explorer planning his or her own journey of discovery.
381 PAGES. 6x9 PAPERBACK. ILLUSTRATED. FOOTNOTES. BIBLIOGRAPHY. INDEX. $16.95. CODE: SAM

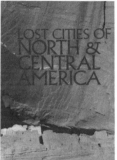

LOST CITIES OF NORTH & CENTRAL AMERICA
by David Hatcher Childress

Down the back roads from coast to coast, maverick archaeologist and adventurer David Hatcher Childress goes deep into unknown America. With this incredible book, you will search for lost Mayan cities and books of gold, discover an ancient canal system in Arizona, climb gigantic pyramids in the Midwest, explore megalithic monuments in New England, and join the astonishing quest for lost cities throughout North America. From the war-torn jungles of Guatemala, Nicaragua and Honduras to the deserts, mountains and fields of Mexico, Canada, and the U.S.A., Childress takes the reader in search of sunken ruins, Viking forts, strange tunnel systems, living dinosaurs, early Chinese explorers, and fantastic lost treasure. Packed with both early and current maps, photos and illustrations.
590 PAGES. 6x9 PAPERBACK. ILLUSTRATED. FOOTNOTES. BIBLIOGRAPHY. INDEX. $16.95. CODE: NCA

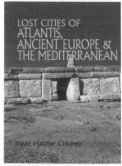

LOST CITIES OF ATLANTIS, ANCIENT EUROPE & THE MEDITERRANEAN
by David Hatcher Childress

Childress takes the reader in search of sunken cities in the Mediterranean; across the Atlas Mountains in search of Atlantean ruins; to remote islands in search of megalithic ruins; to meet living legends and secret societies. From Ireland to Turkey, Morocco to Eastern Europe, and around the remote islands of the Mediterranean and Atlantic, Childress takes the reader on an astonishing quest for mankind's past. Ancient technology, cataclysms, megalithic construction, lost civilizations and devastating wars of the past are all explored in this book.
524 PAGES. 6x9 PAPERBACK. ILLUSTRATED. $16.95. CODE: MED

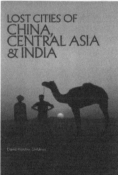

LOST CITIES OF CHINA, CENTRAL ASIA & INDIA
by David Hatcher Childress

Like a real life "Indiana Jones," maverick archaeologist David Childress takes the reader on an incredible adventure across some of the world's oldest and most remote countries in search of lost cities and ancient mysteries. Discover ancient cities in the Gobi Desert; hear fantastic tales of lost continents, vanished civilizations and secret societies bent on ruling the world; visit forgotten monasteries in forbidding snow-capped mountains with strange tunnels to mysterious subterranean cities! A unique combination of far-out exploration and practical travel advice, it will astound and delight the experienced traveler or the armchair voyager.
429 PAGES. 6x9 PAPERBACK. ILLUSTRATED. FOOTNOTES & BIBLIOGRAPHY. $14.95. CODE: CHI

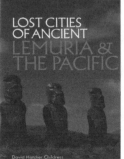

LOST CITIES OF ANCIENT LEMURIA & THE PACIFIC
by David Hatcher Childress

Was there once a continent in the Pacific? Called Lemuria or Pacifica by geologists, Mu or Pan by the mystics, there is now ample mythological, geological and archaeological evidence to "prove" that an advanced and ancient civilization once lived in the central Pacific. Maverick archaeologist and explorer David Hatcher Childress combs the Indian Ocean, Australia and the Pacific in search of the surprising truth about mankind's past. Contains photos of the underwater city on Pohnpei; explanations on how the statues were levitated around Easter Island in a clockwise vortex movement; tales of disappearing islands; Egyptians in Australia; and more.
379 PAGES. 6x9 PAPERBACK. ILLUSTRATED. FOOTNOTES & BIBLIOGRAPHY. $14.95. CODE: LEM

ORDER FORM

10% Discount When You Order 3 or More Items!

One Adventure Place
P.O. Box 74
Kempton, Illinois 60946
United States of America
Tel.: 815-253-6390 • Fax: 815-253-6300
Email: auphq@frontiernet.net
http://www.adventuresunlimitedpress.com

ORDERING INSTRUCTIONS

✓ Remit by USD$ Check, Money Order or Credit Card
✓ Visa, Master Card, Discover & AmEx Accepted
✓ Paypal Payments Can Be Made To:
    info@wexclub.com
✓ Prices May Change Without Notice
✓ 10% Discount for 3 or More Items

SHIPPING CHARGES

United States

✓ POSTAL BOOK RATE

✓ Postal Book Rate { $4.50 First Item
                     50¢ Each Additional Item
✓ Priority Mail { $7.00 First Item
                  $2.00 Each Additional Item
✓ UPS { $9.00 First Item (Minimum 5 Books)
        $1.50 Each Additional Item
    NOTE: UPS Delivery Available to Mainland USA Only

Canada

✓ Postal Air Mail { $19.00 First Item
                    $3.00 Each Additional Item
✓ Personal Checks or Bank Drafts MUST BE
    US$ and Drawn on a US Bank
✓ Canadian Postal Money Orders OK
✓ Payment MUST BE US$

All Other Countries

✓ Sorry, No Surface Delivery!
✓ Postal Air Mail { $19.00 First Item
                    $7.00 Each Additional Item
✓ Checks and Money Orders MUST BE US$
    and Drawn on a US Bank or branch.
✓ Paypal Payments Can Be Made in US$ To:
    info@wexclub.com

SPECIAL NOTES

✓ RETAILERS: Standard Discounts Available
✓ BACKORDERS: We Backorder all Out-of-Stock Items Unless Otherwise Requested
✓ PRO FORMA INVOICES: Available on Request
✓ DVD Return Policy: Replace defective DVDs only

ORDER ONLINE AT: www.adventuresunlimitedpress.com

10% Discount When You Order 3 or More Items!

Please check: ☑

☐ This is my first order   ☐ I have ordered before

Name	
Address	
City	
State/Province	Postal Code
Country	
Phone: Day	Evening
Fax	Email

Item Code	Item Description	Qty	Total

Please check: ☑

☐ Postal-Surface
☐ Postal-Air Mail (Priority in USA)
☐ UPS (Mainland USA only)

Subtotal ▶	
Less Discount-10% for 3 or more items ▶	
Balance ▶	
Illinois Residents 6.25% Sales Tax ▶	
Previous Credit ▶	
Shipping ▶	
Total (check/MO in USD$ only) ▶	

☐ Visa/MasterCard/Discover/American Express

Card Number:

Expiration Date: Security Code:

✓ SEND A CATALOG TO A FRIEND: